MYSTERIES OF

SPAC

The James Webb Space Telescope penetrated the Tarantula Nebula's clouds to capture thousands of stars, including never before seen young ones enveloped in cosmic dust.

CE

EXPLORE THE DEPTHS OF THE UNIVERSE

NATIONAL GEOGRAPHIC
WASHINGTON, D.C.

CONTENTS

INTRODUCTION
4

PART 1
THE MYSTERIES OF SPACE TRAVEL
6

PART 2
THE MYSTERIES OF ALIEN LIFE
36

PART 3
THE MYSTERIES OF DEEP SPACE
64

ILLUSTRATIONS CREDITS
96

Flanked by companion galaxies in this composite image, the Cartwheel galaxy is the result of a 400-million-year-old collision. Its inner and outer rings are still expanding from the impact.

INTRODUCTION

IT OFTEN SEEMS that for every scientific mystery solved, a dozen more unfold. We've sent humans to the moon—but we still don't know how it was formed, exactly how old it is, or whether it could truly be an outpost for human habitation. Our probes have visited the satellites of Jupiter and Saturn, only to confirm that we really need to go back to these icy worlds and their tantalizing possibilities of deep-sea life. Space telescopes such as Hubble have revealed much of the distant universe, but brand-new eyes in the sky are upending our ideas of galaxy formation and the very beginnings of the cosmos itself.

Building on the achievements of the 20th century, scientists and explorers are using innovative technologies to dig into these mysteries as a range of new missions gets under way. Soon, the Artemis program will land astronauts on the moon again, the first boots to walk that dusty surface since 1972. Explorers are swimming through cave waters and probing Arctic ice to discover just what kind of hardy life might exist in such darkness—and whether those sorts of organisms might be found under the ice of worlds such as Europa or Enceladus. First images from the James Webb Space Telescope have already wowed both astronomers and the public with visions of galaxies and nebulae, as well as with key information about exoplanets and their atmospheres—again, possible clues to the existence of extraterrestrial life.

In these pages, we look at these new initiatives and at some of the other tantalizing questions we're asking right now. What, for instance, are rogue planets, and what do they mean for future exploration? How do slow-rolling gravitational waves reveal the origins of the universe? And what does it all mean for our existence here on this warm and watery planet? —PATRICIA S. DANIELS

Astronaut Neil Armstrong and the Apollo 11 *Eagle* moon lander are clearly reflected in Buzz Aldrin's helmet.

LEFT

Dignitaries, including Spiro Agnew, Lady Bird Johnson, and Lyndon Johnson, watch Apollo 11's liftoff at the Kennedy Space Center.

Observers take in the launch of NASA's Space Launch System (SLS) rocket, carrying the Orion spacecraft, during the Artemis I flight test.

PART 1

THE MYSTERIES OF SPACE TRAVEL

In the Artemis program, everything old is new again. The dream of landing men on the moon dates to the 1960s; the reality of sending women and people of color has taken until the 2020s. The immense human ingenuity that goes into it all is the same. And on launch day, the earthbound will always marvel as rocket fire turns into a pinpoint of light in the heavens.

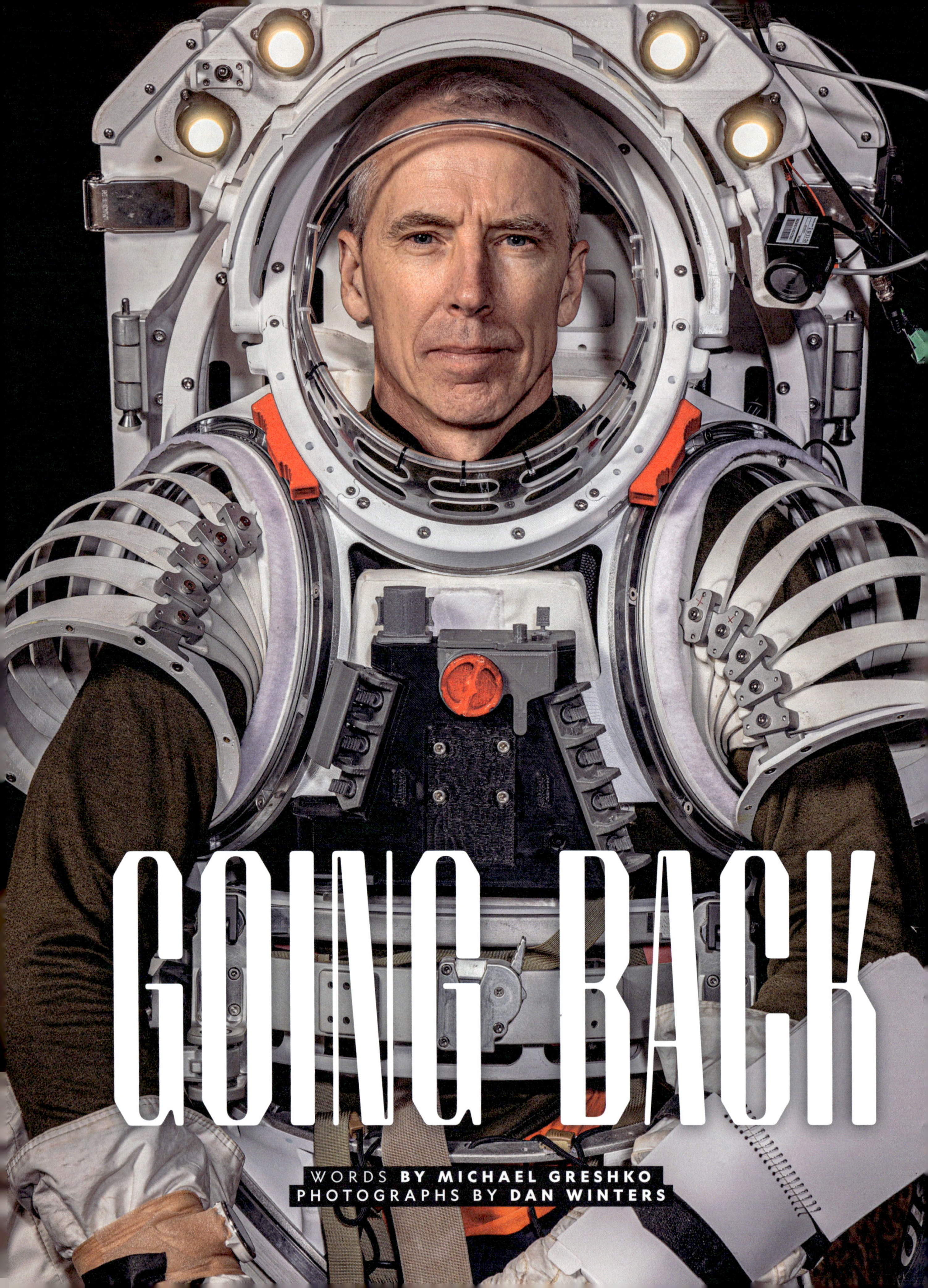

GOING BACK

WORDS **BY MICHAEL GRESHKO**
PHOTOGRAPHS BY **DAN WINTERS**

TO THE MOON

NASA'S ARTEMIS PROGRAM BRINGS US ONE GIANT LEAP CLOSER TO THE MOON, MARS, AND BEYOND.

'I LOVE IT WHEN IT JUST TURNS INTO A STAR!'

CHRISTINA KOCH EXCLAIMED. The NASA astronaut and three colleagues, dressed in blue flight apparel, were standing on a knoll at the Kennedy Space Center in Florida peering into the night sky as the most powerful rocket that had ever launched turned into a pinprick of light.

Minutes earlier, at 1:47 a.m. on November 16, 2022, the 32-story flying machine known as the Space Launch System (SLS) had lifted off. Through my binoculars, the rocket's orange pillar of exhaust was nearly blinding. Every crackle of its 8.8 million pounds of thrust—equivalent to 31 jumbo jets—rattled my lungs.

That colossal rocket, hurtling downrange at more than 17,500 miles an hour, hoisted aloft the Orion spacecraft, designed to carry astronauts farther into space than they have ever ventured. To measure how deep space will affect astronauts, the gumdrop-shaped crew module carried a mannequin named Campos and two female "phantoms," or artificial torsos. Over the subsequent 25 days, 10 hours, and 53 minutes, the test dummies would soar to more than a quarter million miles from Earth before plunging back through the atmosphere at nearly 25,000 miles an hour. The next Orion will have four people on board as it voyages around the moon. Koch (pronounced Cook) hoped to be among them.

The launch of this 2022 mission, named Artemis I, marked a milestone for NASA, which aims to put humans back on the moon for the first time in more than 50 years. If all goes as planned, Artemis II will send a crew on a lunar flyby no sooner than September 2024. Then comes Artemis III—a crewed landing—planned for 2026, followed by more missions to establish a lunar presence.

Why go back to the moon? For one, the lunar surface remains a scientific wonderland. Its rock and dust chronicle the sun's changing activity over 4.5 billion years. Its craters could reveal secrets from the ancient bombardments that also hit Earth. The icy schmutz around the lunar north and south poles might offer insights into how water finds its way through the solar system. Artemis plans to land crews near the south pole to study these suspected frozen water deposits, a step toward possibly harvesting ice for water, oxygen, and rocket fuel.

There are political calculations, too: international cooperation, aerospace contracts, skilled jobs.

Beyond that, the moon is preparation for a crewed journey to Mars, perhaps in the 2030s, as part of the agency's push to find out whether the red planet ever harbored life. The moon and Mars differ, but both are forbidding

The SLS and Orion spacecraft sit in the world's largest single-story structure, the Vehicle Assembly Building at the Kennedy Space Center in Florida.

PREVIOUS PHOTO

When astronauts Drew Feustel and Zena Cardman did practice moonwalks, they each wore a training suit weighing more than 80 pounds to simulate lunar gravity.

Engineers, like NASA contractor Don Biggs, run launch simulations for SLS's intricate flight software on hardware mounted to mimic the inside of the rocket.

"THIS IS TURNING THE FIRST PAGE ON A BRAND-NEW CHAPTER OF SPACE EXPLORATION."

JACOB BLEACHER, NASA

realms where humans need technologies such as pressurized habitats and advanced space suits to survive. And the moon is only a few days away. With the engines we have today, it could take seven to nine months to reach Mars.

Artemis has faced challenges. Years of delays. Billions in cost overages. Skepticism that humans are even needed for space exploration. But if Artemis succeeds, it won't just return astronauts to the lunar surface. It could also begin an era of vast possibility and humbling responsibility: one where humankind regularly lives and works on worlds beyond our own. "This is turning the first page on a brand-new chapter of space exploration," said Jacob Bleacher, NASA's chief exploration scientist.

TESTING THE EQUIPMENT

One clear night in October 2022, astronaut Zena Cardman clambered through a field of lava rock in a mock-up space suit, her eyes locked on the moonlike landscape. Cardman, a geobiologist, and fellow astronaut Drew Feustel, a geophysicist, were on a mission to collect rock samples near Arizona's S P Crater, an 820-foot-tall cinder cone that formed during an ancient volcanic eruption.

Long shadows slithered across the craggy landscape. At a distance, a scientist in a baseball cap wheeled a spotlight around on a small cart, keeping it trained on the two astronauts. When Cardman and Feustel dipped out of the glare—meant to replicate the harsh sunlight at the lunar south pole—they couldn't see more than 30 feet. The duo tried orienting themselves with low-resolution maps, meant to simulate working off satellite images. No GPS or compass bearings were allowed: Neither would be much use on the moon missions.

Astronauts Cardman and Feustel trekked for miles across the San Francisco Volcanic Field in Arizona to test how well Artemis crews will be able to navigate in the harsh light on the moon.

"It was actually remarkably difficult to pinpoint where we were," Cardman said. "Mission Control is telling us, 'We're pretty sure you should see a hill in front of you'... and we're like: 'Maybe!'"

The main goal of this simulated moonwalk and other analog missions isn't to train astronauts. Instead, they're meant to test everything else, from the chisels that moonwalkers will wield to the procedures that Mission Control will follow back on Earth. And then, after years of careful rehearsal, it will be time for liftoff. "It's starting to sink in that it's real," said Jessica Meir, a member of the astronaut corps since 2013. "Nothing of this scale has ever launched during my lifetime."

TO THE MOON

AND BEYOND

Bill Nelson recalls exactly where he was when Apollo 11 commander Neil Armstrong took his first steps on the moon: in a hotel room in London, just before 4 a.m., enraptured by the grainy broadcast on a black-and-white TV. Nelson grew up in Melbourne, Florida, as scientists and rocketeers were turning nearby Cape Canaveral—Apollo's eventual home port—into a missile test site. Later, in his career as a congressman and senator from Florida, he established himself as a legislative leader on space policy. In 1986 he even flew on a six-day space shuttle mission. Now 80 and a NASA administrator, he is overseeing Artemis's first missions.

Artemis as we know it today grew out of a Trump-era push to put humans on the moon as a stepping stone to Mars. The Biden administration has embraced it with minor tweaks. But arguably, the Artemis era began on February 1, 2003, when the space shuttle *Columbia* broke apart during reentry and killed its seven crew members. The disaster, precipitated by mid-launch damage to the shuttle's fragile heat shield, hastened the fleet's retirement—and raised the question of what would follow.

In 2004, the George W. Bush administration unveiled a moon-to-Mars strategy to replace the shuttle, enshrined into law as the Constellation program. The initiative, which included Orion, soon blew its budget and fell behind schedule, and in 2010 the Obama administration proposed canceling it. But lawmakers, co-led by Nelson, voted to keep Orion going and backed two new rocket

DESIGNING WOMAN

Sabrina Thompson is on a mission to empower women, starting with a flight suit that fits.

Since 1963, when Soviet cosmonaut Valentina Tereshkova became the first woman to slip the bonds of Earth, dozens have followed in her contrails. Despite this, flight suits haven't exactly been designed with women in mind. Sabrina Thompson, founder of the brand Girl in Space Club and a NASA engineer, has created a suit for female fliers that, she says, is not only functional but also fashionable.

"I listened to the women who would wear it," Thompson says, "and I'm still listening." Her suit, meant for a variety of environments, including the International Space Station, is "a blend of STEM, art, and fashion." Thompson is also developing a pressurized version that could be worn during launch and reentry. She hopes to start a movement that results in women exploring space in chic comfort—and inspires them to create what they need to succeed, whether on Earth or off.

—ANNIE ROTH

FLATTERING CUSTOM FIT

Each suit can be tailored to accommodate various sizes and shapes. The form-hugging garment follows body contours while allowing the wearer to maintain a full range of motion.

A BIT OF FLAIR

Patches and accent fabrics on areas such as the chest and shoulder “add a touch of elegance and style to the women’s flight suit” without sacrificing functionality, Thompson says.

CONVENIENCE FLAP

With this patent-pending feature, the back of the suit can unzip and snap onto the front, allowing its wearer to take bathroom breaks without having to remove the entire suit.

ADJUSTABLE BELT

The user can tighten or loosen the waistline according to her preference. The belt, which has a quick-release buckle, also prevents any excess fabric from hanging loose and getting caught on equipment.

Thompson has incorporated multiple elements into her design.

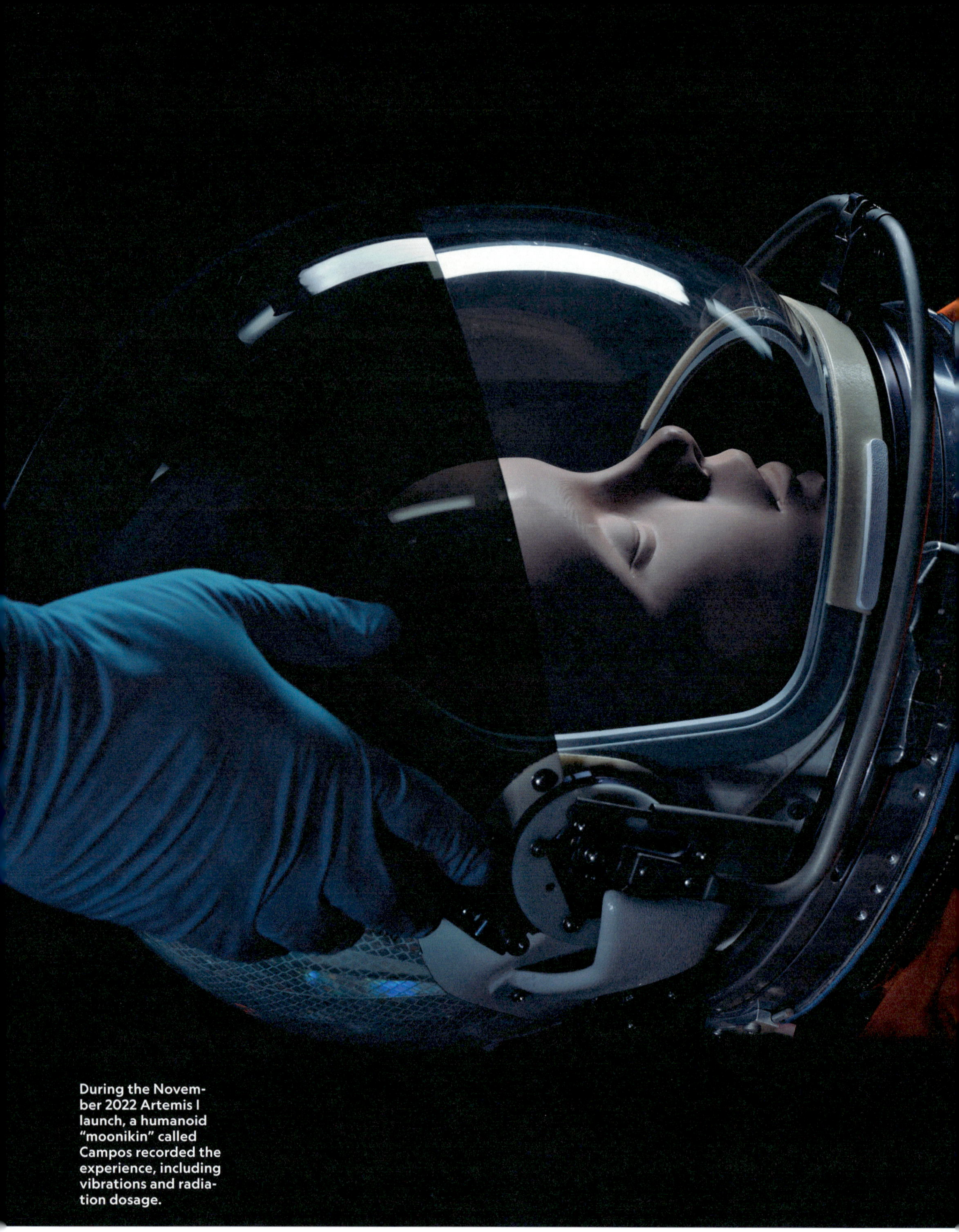

During the November 2022 Artemis I launch, a humanoid "moonikin" called Campos recorded the experience, including vibrations and radiation dosage.

"IT'S STARTING TO SINK IN THAT IT'S REAL...NOTHING OF THIS SCALE HAS EVER LAUNCHED DURING MY LIFETIME."

JESSICA MEIR, ASTRONAUT

strategies: the Commercial Crew program, which contracts with SpaceX and Boeing to ferry astronauts to the International Space Station (ISS); and the SLS, whose design borrowed as much technology as possible from the 1970s-era space shuttle.

All of this represents a substantial national commitment. The NASA Office of Inspector General estimates Artemis's total cost through September 2025 will hit $93 billion. Still, total spending on Apollo exceeded $280 billion in today's dollars, and after adjusting for inflation, Apollo's peak annual cost was roughly 60 percent more than NASA's entire budget today.

THE CAPSULE AND THE ROCKET

For all the superficial similarities to Apollo—the big rocket, the gumdrop capsule—Artemis has modern ambitions. The missions will be longer: Orion was designed to keep a crew of four alive for up to 21 days. And the electronics on board SLS and Orion are more sophisticated. But NASA's newest vehicles still rely on Apollo's manufacturing network.

Both are born in New Orleans 15 miles east of downtown at NASA's Michoud Assembly Facility. For decades the 43-acre plant has played a key role in the agency's rocket manufacturing. Once a French sugar plantation, it was taken over in 1961 by NASA, eager for a site with a deepwater port to build Apollo's Saturn V rocket. The first stage was made here. So was the shuttle's fuel tank.

In April 2023, SpaceX conducted the first full test flight of the reusable rocket known as Starship. A variant of its upper stage will act as Artemis III's lunar lander.

Unlike NASA, SpaceX rapidly builds and tests many prototypes, increasing the risk of "rapid unscheduled disassembly"—such as this fireball in December 2020 (right).

"Until recently, you didn't get out of Earth orbit unless you went through Michoud," said Amanda Gertjejansen, an engineer with Boeing, an SLS prime contractor.

The construction of SLS core stages moves across Michoud from east to west. Curved aluminum panels are welded into cylindrical segments that get stacked and then welded together. After welds are checked, insulation is added, and avionics systems are installed, finished core stages are floated by barge to Florida. Orion also makes its way to the Sunshine State: Following construction of its main metallic skeleton, led by Lockheed Martin, it's shipped to Kennedy Space Center for assembly.

Before that barge ride, the core stage sits in a cavernous room on its side for final assembly. And it was in the room where, safety glasses on, Gertjejansen led me around the rocket. At 27.6 feet in diameter and 212 feet long when fully assembled, it felt preposterously big: a pharaonic monument awaiting a final lift into place. Silvery piping ran along the outside to feed fuel to the engines. When we poked our heads into the engine section—the part of the rocket she manages—the plumbing-stuffed cavity had the feel of a cramped jungle gym.

As sophisticated as this behemoth is, artisans also contributed. Some of the foam insulation is hand-sprayed. Chessboard-like markers along its sides—used to track orientation and speed—are hand-painted. "When you think of building a rocket, you think of rocket scientists," Gertjejansen told me. "Those are only a very small part."

TROUBLESHOOTERS ON ALERT

At NASA's Marshall Space Flight Center in Huntsville, Alabama, a wall of TV monitors features live video feeds, from every conceivable angle, of Launch Complex 39B, the Artemis launchpad. During tests and launch attempts, this room holds a team of experienced SLS engineers, ready to troubleshoot a problem. "You put plans in place to deal with the things that you don't plan on, so when they happen, you're not reacting—you're responding," said David Beaman, who manages this team and hundreds of other engineers.

According to Beaman, NASA's current safety approach is less susceptible to human error: It's a strategy based on taking the extra time to make sure every nagging concern is resolved. "If it means making a few hard decisions, and waiting a few more days, and getting a little more data, we're gonna do that," he told me.

NASA scrubbed Artemis I's first launch attempt in August 2022 amid concerns over weather, a misbehaving temperature sensor, and a hydrogen fuel leak. More leaks and several hurricanes delayed liftoff until fall. Hours

SPACE SUIT BARBIE

For 60 years, Barbie has inspired girls to reach for the stars.

She may be a toy, but Astronaut Barbie has been a role model since 1965, well before American women were admitted to the space program. In 2021, the iconic doll became a space suit tester. Undergraduates at Washington State University won the Artemis Award for their research into novel ways of cleaning abrasive lunar dust from space suits, using liquid nitrogen spray. Their test subject? A Barbie, clad in a tailor-made suit and covered with ash from Mount St. Helens. (It's illegal to own actual moondust.) The nitrogen worked well in picking up and containing the dust. Astronaut Barbie was unperturbed by the whole process. Her role in the research has now ended, but the scientists still bring her to conferences as a small but stylish exhibit. —PATRICIA S. DANIELS

1965 Miss Astronaut Barbie wore a Mercury-style suit, inviting dreams of space travel.

before its November 16 launch, the Orion rocket was almost fully fueled—and dangerous to approach. NASA sent a "red crew" out to tighten nuts on a valve. "The rocket, it's alive. It's creaking. It's making venting noises. It's pretty scary," said Trenton Annis, who helped make the repair.

On November 28, 2022, Orion flew beyond the moon to more than 268,500 miles away—nearly 20,000 miles farther than any other "human rated" round-trip mission—and then on December 11 plunked into the ocean off Mexico's Baja California peninsula.

THE CREW

COMES TOGETHER

Artemis II is a go, a 10-day sojourn around the moon that will come within 6,500 miles of the lunar surface. Koch, an electrical engineer and one of Artemis II's mission specialists, received the historic assignment in March 2023. Three military pilots will be joining her: NASA's Reid Wiseman and Victor Glover and the Canadian Space Agency's Jeremy Hansen.

Wiseman, Artemis II's commander, served as NASA's chief astronaut—who chooses crews but can't join them—until two days before Artemis I launched. He regained his flight eligibility just in time. Glover, Artemis II's pilot, was the second-in-command on NASA's SpaceX Crew-1 mission to the ISS, during which he spent 168 days in orbit. Hansen, Koch's fellow mission specialist, was the first Canadian to supervise the training of a NASA astronaut class. Artemis II will be his first spaceflight. Glover will be the first Black astronaut to venture beyond low Earth orbit. Koch and Hansen will be the first woman and non-U.S. citizen, respectively, to do the same.

Preparations for the mission are well under way. In a bright room in Kennedy Space Center, technicians in white "bunny suits" have been testing and assembling the mission's Orion spacecraft, with the Artemis III Orion taking shape nearby. A team led by astronaut Stan Love has been training to serve as Artemis II's capcoms, the main points of contact between the crew and Mission Control, ready to relay technical directions or detect the quaver in an exhausted astronaut's voice.

Ultimately, NASA plans to build a small space station called Gateway to act as a staging ground for weeks-long missions to a lunar outpost with habitats, power stations, landing pads, and pressurized rovers—all of which are still in the concept stage.

This immense infrastructure would be too much for the United States to build alone, so Artemis is teaming

1985
Astronaut Barbie's fashionable suit sported '80s-style shoulders and matching boots.

2013
Mars Explorer Barbie was decked out for visiting the red planet, complete with pack.

2021
Space Discovery Barbie was the first to travel to the International Space Station.

up with foreign partners. For more than 22 years, the ISS has orbited 250 miles above our heads with humans on board, thanks to collaboration among the United States, Russia, Canada, Europe, and Japan. This union has spread the station's costs around and made canceling the program diplomatically difficult. NASA has drawn on those lessons for Artemis: The European Space Agency (ESA) provides Orion's service module, which contains the spacecraft's solar arrays and main engine. That cooperation also extends to Gateway, which NASA is building with ESA and the space agencies of Canada and Japan.

NASA astronauts are acutely aware that they are not only public figures but also inspirational symbols: Symbols of exploration, of science, of the national spirit. Koch has wrestled with how to handle these expectations. For 328 days, she lived and worked aboard the ISS, setting a record for the longest spaceflight

The crew of the Artemis II mission (clockwise from top left): Mission Commander Reid Wiseman, Specialist II Jeremy Hansen, Pilot Victor Glover, and Specialist I Christina Koch.

by a woman. Koch's initial impulse was to downplay the milestone. Then a former colleague reminded her that her achievement might give people a sense of greater possibilities.

"How awesome it is that as a species, as humanity, we are undertaking this right now—that we have decided that it's that important," Koch said. "It's because we love exploration. It's because we believe in the power of learning."

SOLAR ENERGY FROM SPACE

The future is near: energy anytime, anywhere.

Though it was fantasy just a few decades ago, space-based solar power may soon be reality—and a renewable energy breakthrough: accessible anytime, unlike ground-based solar panels limited by darkness, and available anywhere on Earth, even if infrastructure is lacking.

How would it work? Enormous space-based structures would support photovoltaic panels that collect sunlight and convert it into radio signals for transmission to Earth. This type of space mission is "one of the most complex engineering problems that we have in front of us," says Mamatha Maheshwarappa, a payload systems lead at the UK Space Agency, which is helping fund project research and development.

The mission also will be pricey—tens of billions of dollars. At Caltech, researchers recently reported significant progress with their Space Solar Power Demonstrator, in orbit since January 2023, testing materials and electronics. It has successfully transmitted wireless energy in space for the first time.

—LIZ KRUESI

The SLS's 177-foot-tall side boosters are filled with solid polymer fuel. The boosters chewed through several tons of the propellant each second during this 2022 test firing in Utah.

GOING THE DISTANCE

On the Artemis I mission, the uncrewed Orion reached its maximum distance from Earth: 268,563 miles, a record for any round-trip mission with a spacecraft built to hold crew. Now, for Artemis II, the SLS must safely carry four humans on a similarly ambitious trajectory as they test its evolvable technology for future explorations, including human missions to the moon and Mars and robotic missions to the moon, Mars, Saturn, and Jupiter.

FOUR RS-25 ENGINES

Together they provide two million pounds of thrust for launch and ascent, accelerating the rocket to 17,000 miles an hour over the first eight minutes of flight and ascending 500 feet in the first seven seconds.

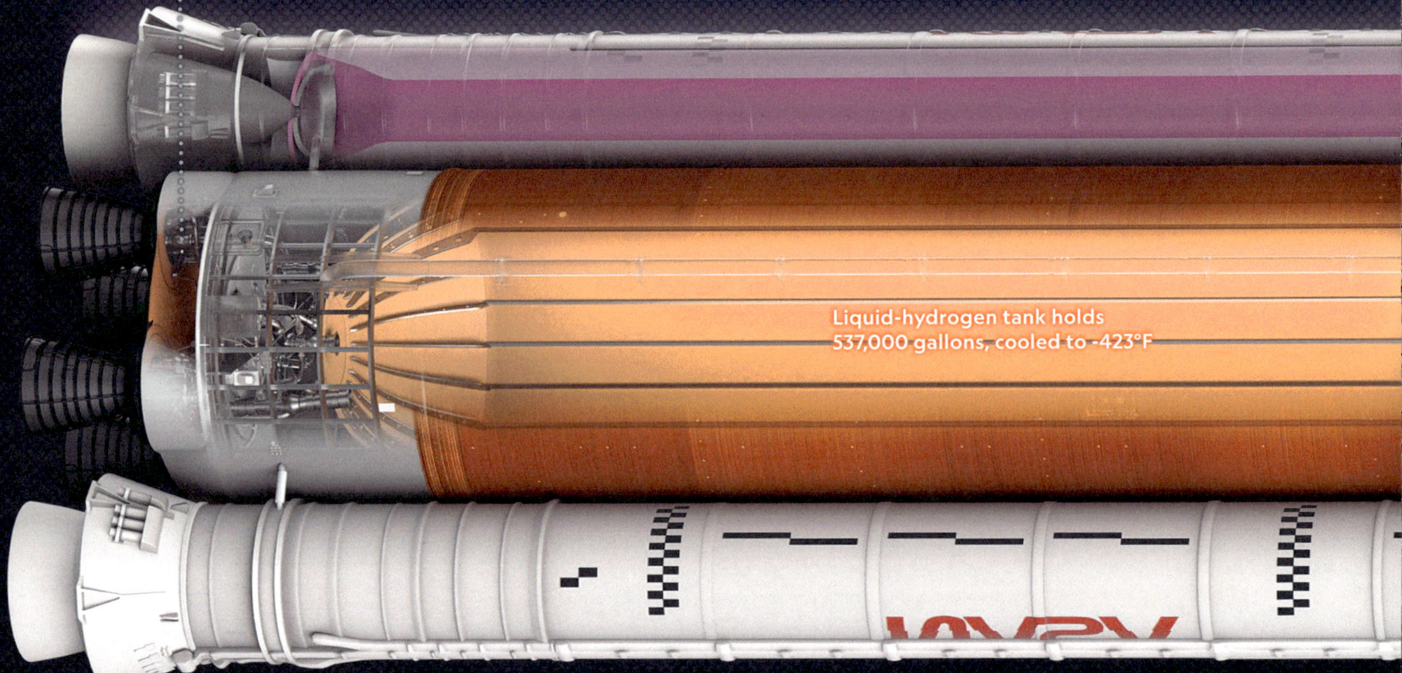

LAUNCH SYSTEMS

The SLS thrust and payload capacities illustrate NASA's progress over time. The agency is also partnering with SpaceX on Starship technology for Artemis moon missions.

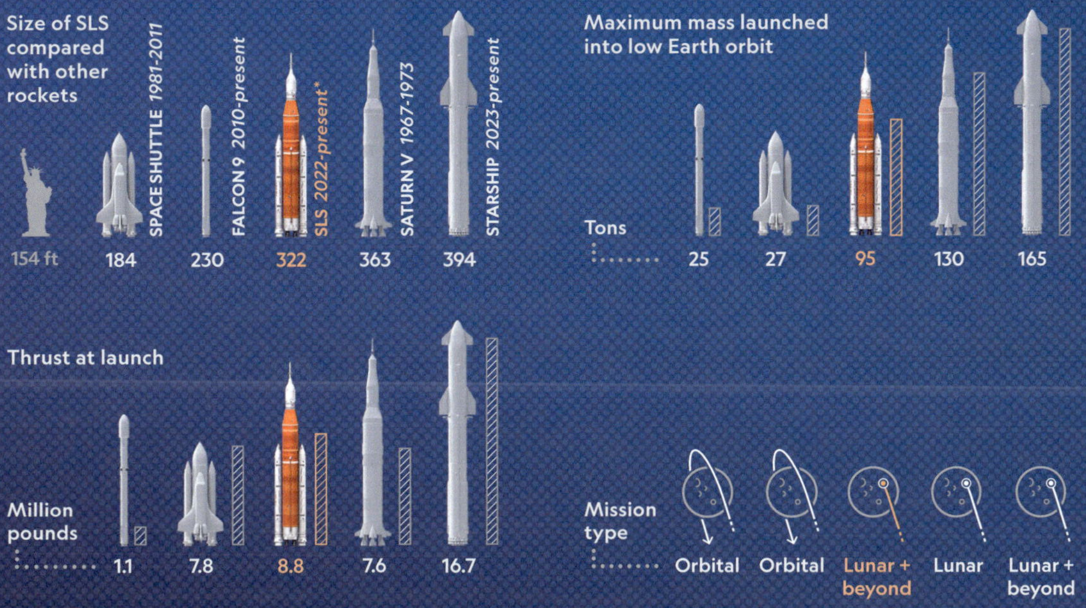

SOLID ROCKET BOOSTER

Standing 17 stories tall and burning some six tons of propellant per second, each booster generates more thrust than 14 four-engine jumbo commercial airliners. The boosters provide 75 percent of total thrust during the first two minutes, 12 seconds of flight.

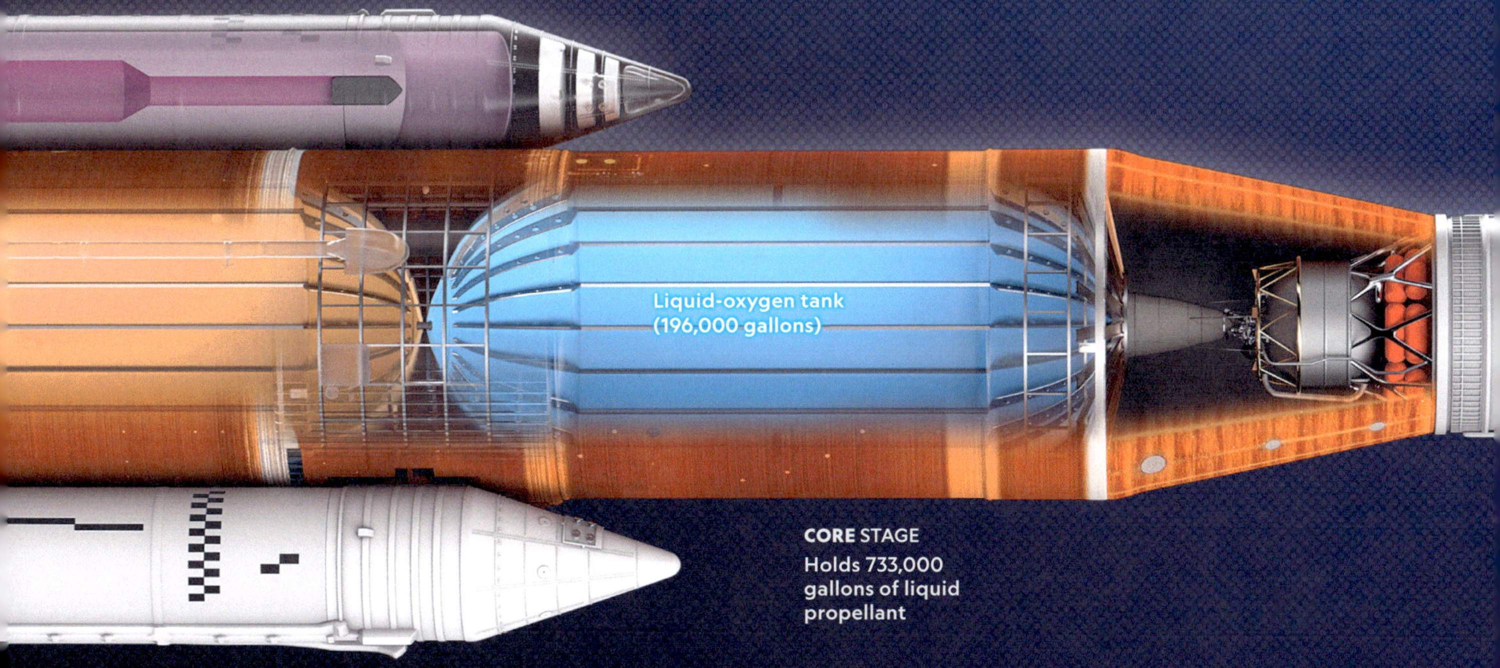

CORE STAGE

Holds 733,000 gallons of liquid propellant

ALBERTO LUCAS LÓPEZ, NGM STAFF; KELSEY NOWAKOWSKI; RONALD PANIAGUA. SOURCE: NASA

SURVIVAL SUIT

The Orion crew survival system (OCSS) suit for launch, ascent, and reentry protects the wearer from hazards and emergencies.

Touch-screen gloves

Layer for pressurized gas

Structural layer for shape

The OCSS suit includes flotation pouches for use as needed during water landings.

Deployed for life raft entry

Upright flotation

Emergency oxygen provides 35 minutes of air in orbit, 10 on Earth.

Lightweight, fire-retardant shoes fit comfortably when pressurized or unpressurized.

The launch vehicle stage adapter also protects electronics from extreme vibration

Orion stage adapter

Rotated and enlarged above

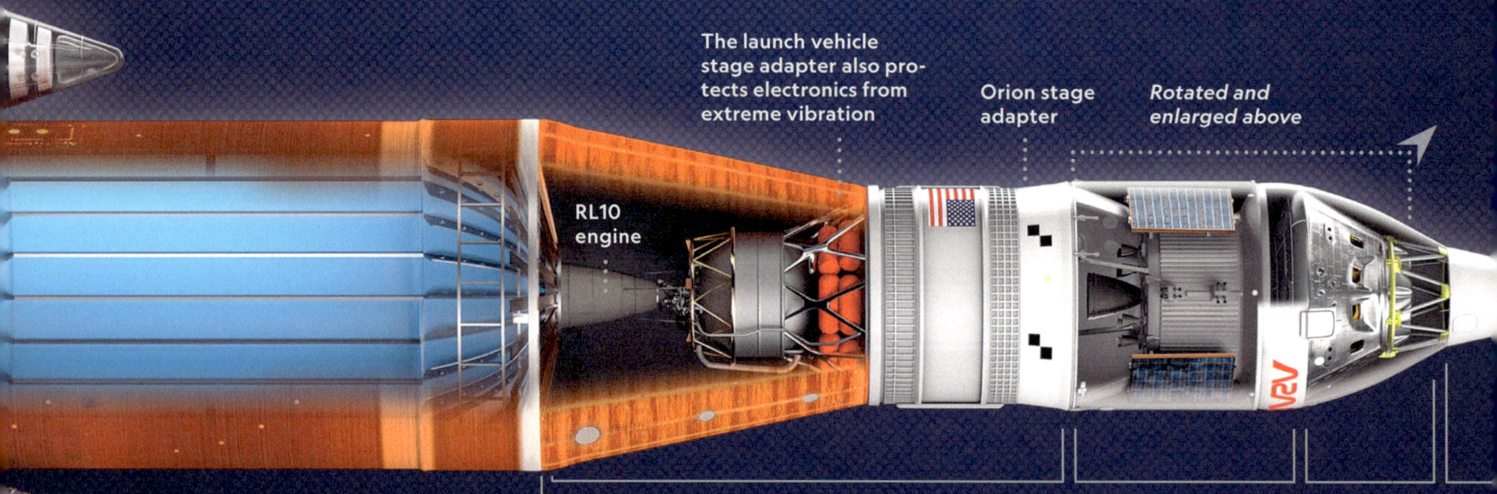

INTERIM CRYOGENIC PROPULSION STAGE (ICPS)
Includes an engine providing 24,750 pounds of thrust

SERVICE MODULE
Source of solar power, thermal control, air and water, and propulsion

CREW MODULE
Pressurized capsule for crew living quarters

Rocket **SPACE LAUNCH SYSTEM**

ALBERTO LUCAS LÓPEZ, NGM STAFF; KELSEY NOWAKOWSKI; RONALD PANIAGUA. SOURCE: NASA

CREW MODULE

For 10 days four astronauts will live within 316 cubic feet, about the size of a single-car garage. They're expected to lose all contact with Earth for up to 45 minutes when the craft flies behind the moon.

Some 18 minutes after launch, solar arrays span out to power the craft.

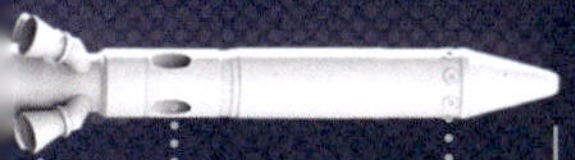

LAUNCH ABORT SYSTEM

Carries crew away from the rocket in case of emergency during launch or ascent

Spacecraft **ORION**

MOON ROCKET

Upon launching Orion into orbit, about eight minutes after takeoff the booster rocket's core stage will detach and fall into the Atlantic Ocean. The Orion crew module, the only reusable part of the rocket, will eventually land in the Pacific Ocean.

HOW WE COULD LIVE ON THE MOON

WORDS BY
PATRICIA S. DANIELS

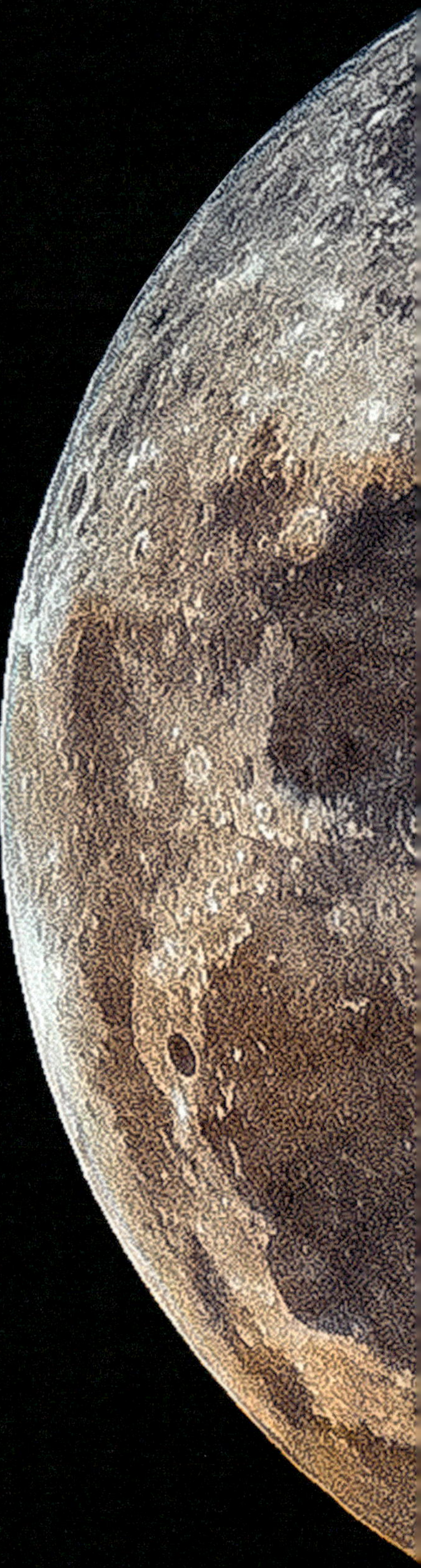

LUNAR BASES ARE PART OF THE PLAN, AND MOON HOUSING COULD BE A REALITY BY 2035.

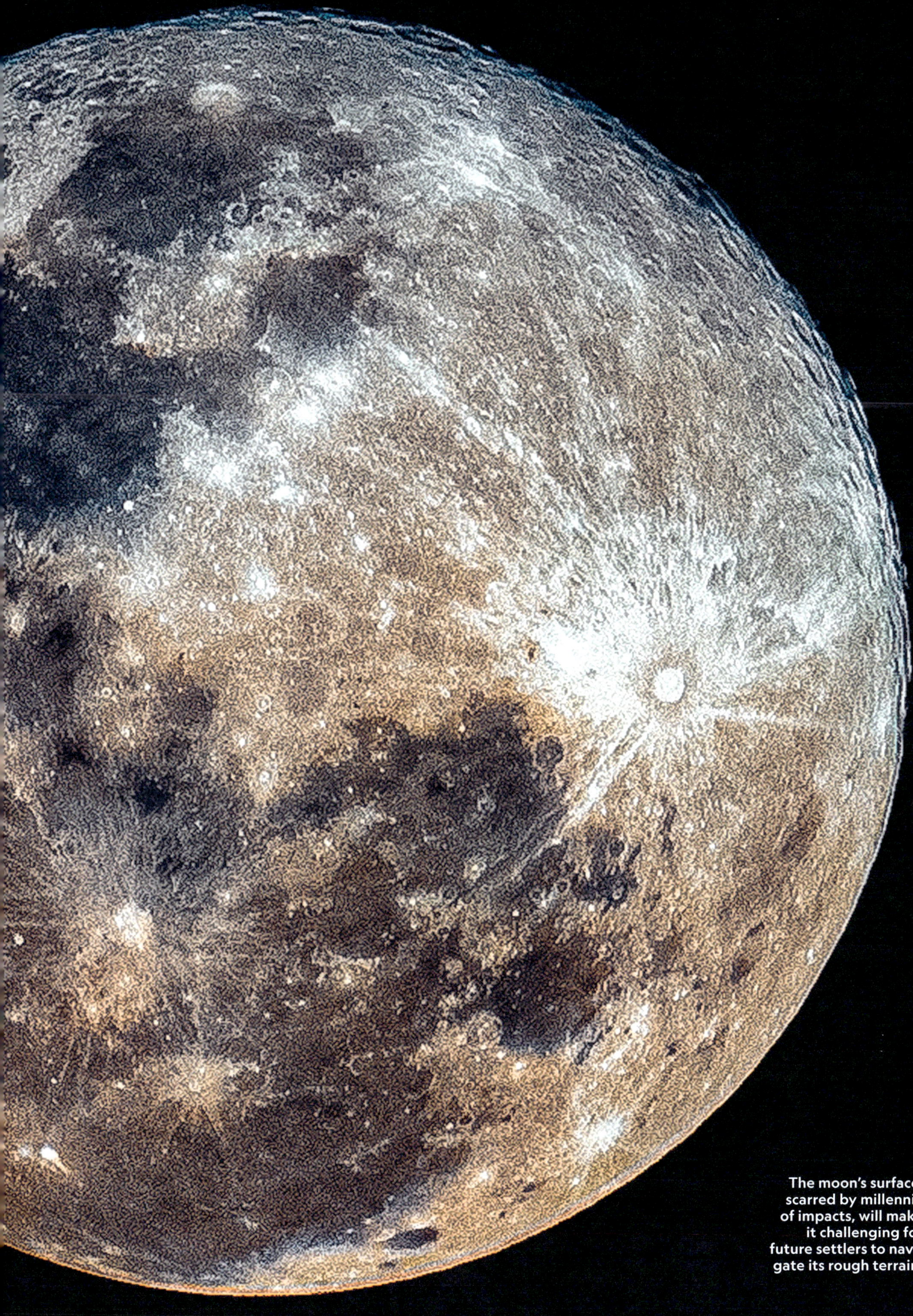

The moon's surface, scarred by millennia of impacts, will make it challenging for future settlers to navigate its rough terrain.

ORBITING THE moon is exciting, landing on the surface even more so, but for the full sci-fi experience, living on the moon takes the prize. And a lunar base—or multiple bases—is indeed part of the long-term Artemis program. It won't happen soon, but starting somewhere around the Artemis VII, VIII, or IX missions in the 2030s, NASA and its partners plan to build at least one structure that will house crews for up to two months at a time.

Such a base will give lunar visitors and their robotic companions the time and elbow room to conduct experiments, explore the surroundings, and assess local resources. Located next to Shackleton crater at the lunar south pole, it could take advantage of the likely wealth of water ice in the crater's shadowed depths, while drawing power from the low-circling sun that shines on the ridges most of the year. It would need to keep its distance, though, from any landing pad for the lunar landers. Those spacecraft will kick up the moon's gritty surface, blasting out high-speed sprays of up to a million pounds of sharp-edged soil and ice particles.

MASS = MONEY

JUST HOW THE BASE WILL BE constructed is still being studied, but one thing is clear: We won't want to ship heavy building materials from Earth to the moon. In space travel, mass equals money. Artemis planners are currently looking at either an inflatable shell on a movable frame or a brick structure made from the lunar regolith itself, with the bricks possibly shaped by 3D printers and laid by robots.

Any such shell would need to protect its inhabitants from the unique hazards of lunar living. On Earth, our atmosphere and magnetosphere protect humans from cosmic radiation, but no such shields guard the moon. Unsheltered astronauts could be exposed to up to 1,000 times the radiation they would face on Earth. Meanwhile, micrometeoroids constantly ping the lunar surface. Traveling at speeds over 1,000 miles an hour, even projectiles the size of a grain of sand could do serious damage to a lunar structure. The base will need to be tough.

It will also need power and communications. Given a location near the rim of Shackleton crater, an area bathed in near-constant sunlight, solar power seems to be the best option most of the time. NASA has already contracted with several companies to develop a suitable solar power array, like with panels held atop tall masts to avoid the shadows cast by the moon's sharp-edged ridges. Like the base, the panels will have to be tough enough to withstand abrasive lunar particles, lightweight enough to fit onto spacecraft, and stable enough to remain upright on the uneven terrain. For times when the base is in darkness, a small nuclear fission station might provide backup power—or even a fusion station, should that holy grail of physics ever materialize.

Communicating to and from Earth, and back and forth across the lunar surface, will require a new system of satellites and modules, still in the planning stages. The ESA is working on the satellite end, while NASA is developing LunaNet, basically internet for the moon. These systems will need to provide rapid and radiation-proof communications for not only data and conversations but also navigation and space weather forecasts. Knowing when the next solar storm is coming could mean the difference between life and death in space and on the moon's surface.

OUT FOR A DRIVE

FURTHER ALONG IN DEVELOPMENT are the vehicles that will take the astronauts out on their missions of exploration and research. The Lunar Terrain Vehicle, an open-top moon buggy, would be similar to the lunar rovers used by Apollo astronauts, though vastly updated. It could be driven by two space-suited astronauts or be operated remotely from Earth, avoiding rocks and craters with its advanced sensors. For a more luxurious travel experience: a pressurized lunar RV, known in NASA-speak as a habitable mobility platform. This roomy vehicle would trundle astronauts around in the comfort of their shirtsleeves for up to two weeks at a time.

Much can change in the decade or more before this kind of a moon base is ready. Even so, NASA and its sister programs are looking even farther afield. Lessons learned on a lunar base will be applied to the complex task of building habitats on Mars. The Martian environment, although different from the moon, has many of the same challenges.

MUCH CAN CHANGE IN THE DECADE OR MORE BEFORE SUCH A MOON BASE IS READY.

NASA's space exploration vehicle, a rover that can house two people for two weeks, tests its mettle in the Arizona desert.

Beneath its frozen crust, Saturn's icy moon Enceladus harbors a liquid global ocean that could contain the ingredients for life.

PART 2

THE MYSTERIES OF ALIEN LIFE

Our notions of alien life have evolved from the old movies with bug-eyed saucer pilots from Mars or Venus. Intrigued by exploration of alien moons, we wonder if life exists in such harsh environments and what that life might look like. Where better to search for analogues than in Earth's extremes? In Norway's ice and Italy's underwater caves, scientists are doing just that.

Lit by snowmobile headlights, a photographer flies a drone across the Norwegian landscape to capture a dome-shaped landform called a pingo.

THE HUNT FOR ALIENS ON EARTH

WORDS BY
NADIA DRAKE

A

AS I THUMB THE THROTTLE

ON MY SNOWMOBILE, IT SKITTERS and glides across a sea of snow and ice. In the dusky twilight, the landscape is painted in shades of ethereal blue. I've spent the day crisscrossing a frozen fjord, one of many in Norway's Svalbard archipelago, a cluster of mountainous islands in the high Arctic where auroras often dance overhead and narwhals, belugas, and walruses patrol the seas.

It's March, and the sun has finally returned to the sky here about a month ago after winter's total darkness. I'm with a half dozen scientists who are hunting peculiar landforms called pingos—or, more specifically, the microbes that live within them. Anchored in permafrost, these domes range in size from mounds to small hills; they seasonally expand and contract as water seeping through them freezes and thaws. They're like an icy eruption in slow motion.

Temperatures hover around minus 15°F as the bundled-up, rifle-toting scientists make multiple trips each day to their study sites, where they collect ice cores and water samples while keeping an eye out for polar bears.

The microbes populating the pingos could offer a glimpse at how alien life might survive on other worlds in the solar system—for instance, icy moons with global seas tucked beneath frozen crusts. That's because, in winter, life inside the pingo "doesn't rely at all on solar energy—it's only using chemical energy," says

Scientists take an ice core from one of Svalbard's pingos. Microbes living in the pingos might use life strategies similar to life-forms on an icy moon.

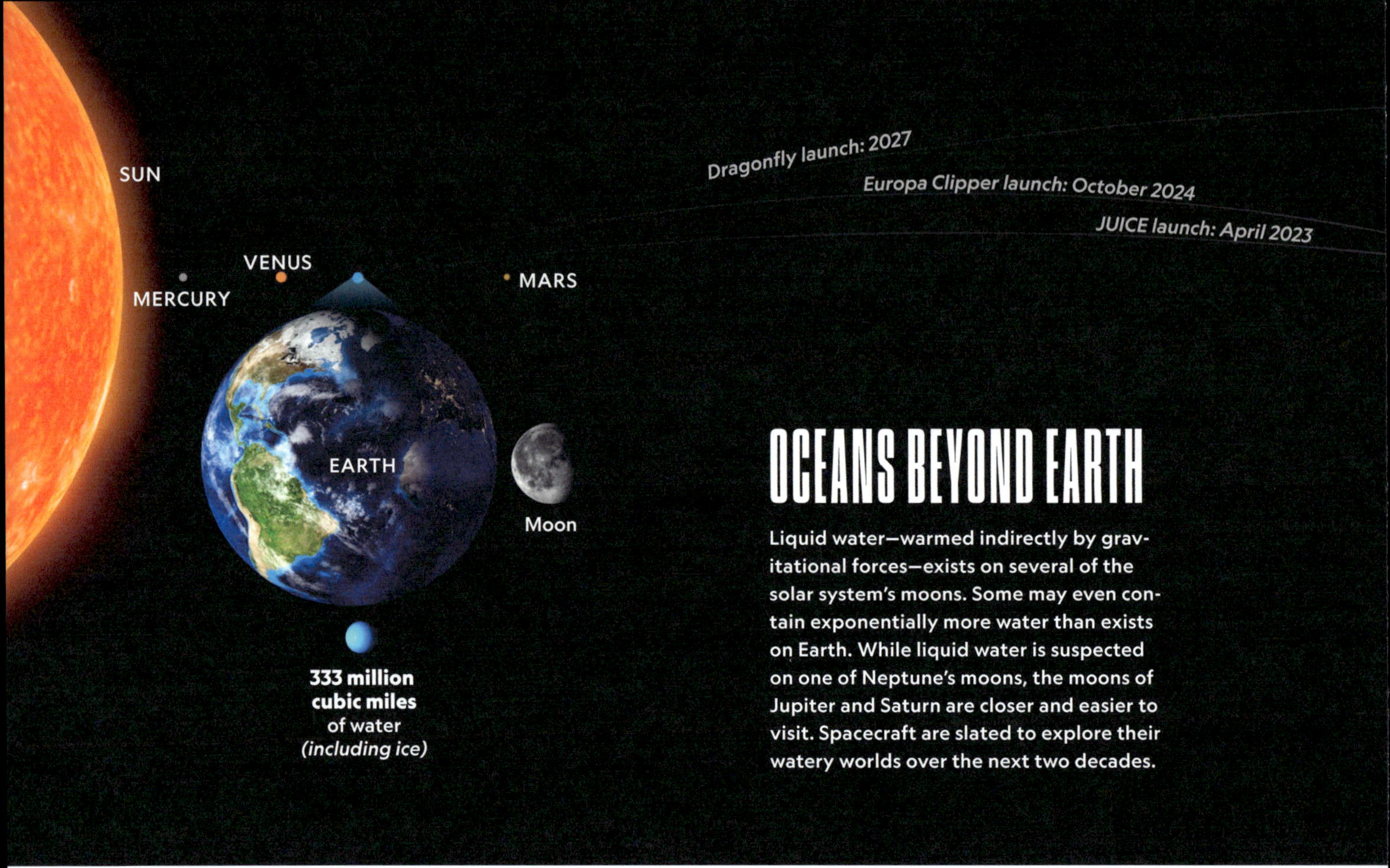

OCEANS BEYOND EARTH

Liquid water—warmed indirectly by gravitational forces—exists on several of the solar system's moons. Some may even contain exponentially more water than exists on Earth. While liquid water is suspected on one of Neptune's moons, the moons of Jupiter and Saturn are closer and easier to visit. Spacecraft are slated to explore their watery worlds over the next two decades.

microbiologist Dimitri Kalenitchenko of UiT The Arctic University of Norway, Tromsø, who is leading the project.

The understanding of sunlight-starved life on Earth is relatively new. For a long time, "we still thought life on this planet was largely restricted to the surface . . . and entirely dependent on photosynthesis," says Barbara Sherwood Lollar, a geologist at the University of Toronto who studies microbes living deep within the Earth. Then, in the late 1970s, the *Alvin* submersible explored a dark oceanic hydrothermal vent near the Galápagos Islands, discovering a flourishing ecosystem about a mile and a half beneath the ocean's surface, and forever changing our conceptions of life's limits. "It's one of the things that forces us to be humble," Sherwood Lollar says. "To think that even on our own planet we are still finding processes, still finding environments that we didn't know existed."

Similarly, scientists used to think a world's habitability depended on its distance from the sun. But that picture is incomplete. Now three faraway moons, warmed by the gravitational tug and pull of the giant planets they orbit, tempt scientists with their promises of alien life in their oceans: Jupiter's moon Europa, where a salty sea containing more water than Earth's oceans sloshes beneath an ice shell; and Saturn's moons Enceladus—a small, ice-encrusted world with a global ocean that erupts through fractures in its south pole—and Titan, with its unearthly terrain of liquid hydrocarbon lakes on its surface and a salty ocean nestled within its interior. Observations suggest that each moon has the chemistry, water, and energy needed to support life as we know it—and maybe don't know it.

We might soon learn if these seas are inhabited. A European Space Agency spacecraft called JUICE is on its way to the Jovian system to surveil the giant planet and its icy moons, including Europa. This year, NASA's Europa Clipper spacecraft will also set sail for Europa, on a quest to solve the mysteries of its crust and salty sea. And later this decade, the U.S. space agency's Dragonfly mission will send an octocopter to Titan, carrying a suite of instruments that could detect signs of life on the hazy moon's surface. Missions to Enceladus are being planned, too.

To prepare for astrobiological exploration so far afield, scientists are testing their instruments and techniques in the cold, dark corners of our own backyard. From Earth's surface to its subterranean caverns, these investigations study some of the most alien creatures on our planet. Their discoveries could even write the story of life's beginnings here, and maybe elsewhere.

"It's a really exciting time to be a planetary scientist," says Morgan Cable of NASA's Jet Propulsion Laboratory (JPL). "We could, for the first time in human history, find life somewhere else."

LIFE UNDER ICE

I first stepped into the Svalbard snow in 2019, as I prepared to board the Norwegian icebreaker *Kronprins Haakon* along with three dozen scientists and engineers. We launched from Longyearbyen and sailed toward a fractured patch of seafloor north of Greenland. There, some two and a half miles down on the ocean floor, the Earth heaves searing dark fluids into the sea, forming what is known as the Aurora hydrothermal vent field. Salt water mingling with hot rocks beneath the seafloor powers these eruptions, and, in turn, these erruptions produce the heat and chemistry needed for organisms to thrive.

"They can host all kinds of weird and wonderful life-forms," said Chris German, a marine geochemist from the Woods Hole Oceanographic Institution who has spent nearly four decades hunting hydrothermal vents, as we cruised north. Under permanent ice cover, the field could be a superlative earthly analogue for the seafloors of Europa and Enceladus.

To explore Aurora's under-ice ecosystems, German and his colleagues brought one of the most advanced uncrewed submersibles on Earth: a three-million-dollar, orange, minivan-size vehicle called *NUI,* short for *Nereid Under Ice.* It can dive three miles down, swim more than 25 miles, and operate for half a day without recharging. *NUI* functions autonomously, but scientists can also direct it to collect particular sediments or organisms.

For days, thick ice—too bulky for the ship to break—had slowed our journey to the vents. Although the going was tricky, the Arctic Ocean's relatively thin icy rind is a friendlier version of the crusts encasing alien oceans. The shell at Enceladus may be less than a mile thick at the south pole, where the geysers erupt, but estimates for Europa's rind suggest it's considerably thicker. Characterizing Europa's frozen shell and the compounds on its surface is one of the highest

Explorers prepare to dive in Lago Verde, in Italy's Frasassi cave system. Its sunless ecosystems could offer a glimpse into alien seas.

priorities for both the Europa Clipper and JUICE missions. Those spacecraft will make multiple flybys of Europa to measure the ice shell's thickness and study its layers and ocean beneath, in hopes that with more information, future missions can get through the rind and access the water.

EXPLORING

THE VENT

When NASA's Cassini spacecraft zoomed around the Saturn system from 2004 to 2017, it sampled the Enceladian plume multiple times and found salts, silica, organic molecules, and molecular hydrogen, all telltale signs of seafloor activity. It also detected phosphorus, an element that's crucial for life on Earth. *NUI*'s job on this cruise was to get the first good look at the Aurora vent field and help figure out whether it's fueling chemical reactions that can support abyssal life-forms—perhaps the kind of life that might evolve in dark, permanently ice-covered alien seas.

Two days after arriving at the Aurora seamount, *NUI* dived to Aurora. As the orange sub sank in the long Arctic twilight, a gentle snow fell on fields of frost flowers carpeting the ice. It took hours for the submersible to descend to the seafloor. In the control room a pilot maneuvered the vehicle. At one point he handed me the joystick, and for a few moments we joked that I held the record for the deepest Arctic remotely operated vehicle dive piloted by a woman. (It was 9:19 a.m., and the sub was at 9,104 feet, but who's keeping track?)

Geomicrobiologists Dani Buchheister (at left) and Jennifer Macalady process samples of a mysterious microbial biofilm collected in Lago Verde.

An hour later, the trouble started. As *NUI* neared its destination, its onboard systems blinked off one by one. Then the pilot reported that he'd lost control of the submersible. After a bit, the team commanded the vehicle to drop its dive weights and begin rising to the surface. Instead, *NUI* sank—to the seafloor.

Three days went by before good news: *NUI* was on its way back up. A corrodible wire attaching the dive weights to the submersible had finally succumbed to the cold, salty seawater that gnawed at it. *NUI* was finally back on board, having helpfully surfaced under a rare patch of thin ice rather than the thick floes covering the area.

NUI wouldn't explore Aurora during this cruise, but it turned out that the vents were still within reach. That same evening, the ice drift cooperated, and the ship sailed right over the vent field, towing a high-tech camera just above the seafloor. Aurora, the imagery revealed, was home to a massive black smoker, a rupture more than five feet across that hurled hot, sulfidic minerals into the sea.

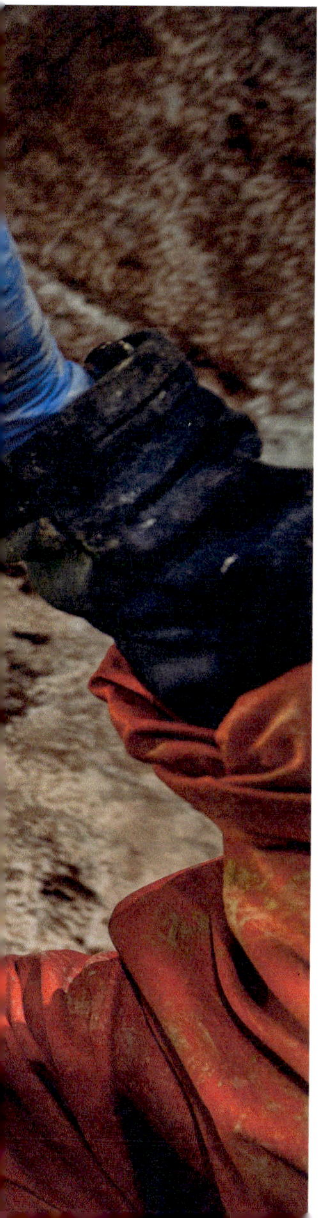

"IF YOU SEE ALIEN CAVE GOO, YOUR FIRST INSTINCT IS TO TURN AROUND."

JENNIFER MACALADY, GEOMICROBIOLOGIST

But as far as hydrothermal vents go, according to the cruise's co-chief scientist Eva Ramirez-Llodra, Aurora seemed remarkably unpopulated. The only organisms in abundance down there were glass sponges, which are sometimes said to be barely alive. Their skeletons are made mostly of silica rather than calcium carbonate. And that makes sense. In the deep sea, silica is common. Life—these otherworldly sponges—found a way.

MYSTERIOUS MICROBES

In February 2023 I visited central Italy's Frasassi cave system with scientists searching for some of Earth's least known creatures: microbes that live deep within the cave's underwater passages. Growing in toxic, oxygen-starved waters, these improbable communities are powered by the slow burn of water interacting with the rock itself, producing metabolic fuels such as hydrogen sulfide and methane—just like the organisms that depend on oceanic hydrothermal vents.

Scientists think that these chemical reactions could exist on icy moons. They also suspect that Frasassi's chemistry is similar to Earth's ancient oceans, where the seeds of terrestrial biology might have sprouted. "Earth was a very different planet when it was born," said Jennifer Macalady, a geomicrobiologist at Penn State University who has been visiting the cave system for more than 20 years. "If we think this aquifer was like Earth's early oceans, and we think that Earth's early oceans might have some similarities with oceans on other planets, this is a great place to hone our skills for life detection."

The system's largest chamber was discovered in the 1970s and is 65 stories tall. It's so big that the cathedral of Milan, an important unit of measurement for Italians, would fit inside. But the only gargoyles here have been sculpted, one drop at a time, by the slow seepage of mineral-rich water. Glistening, slimy, and sparkling with crystals, some of the stalagmites are 65 feet tall and as big around as old-growth redwood trees.

No one knows the extent of Frasassi's network of lakes, which are connected by a vast subterranean aquifer. "There hasn't been much diving in the Frasassi aquifer because, one, it's toxic. Two, it's hard to obtain access.

The Nancy Grace Roman Space Telescope's view will be 100 times wider than Hubble's.

ROGUE WORLDS

Planets without homes roam our universe in much greater numbers than anyone thought.

They are called rogue or free-floating planets, but a better name might be orphaned worlds. For decades, astronomers have known that isolated planets, untethered from any star, are floating throughout our galaxy.

Ejected from their solar systems during the pinballing early days of planetary formation, these planets are dark, so spotting them with conventional telescopes is tough. They are best detected through microlensing—a technique that finds locations in space where the gravity of an unseen mass has distorted the light around it. The first such worlds detected were usually large and rare.

Now, with new findings from New Zealand's Mount John Observatory, we know that rogue planets vastly outnumber sun-tethered worlds like our own by a factor of six. In fact, trillions of dark planets probably drift throughout the Milky Way.

Most of the free-floating planets are smaller worlds, more like Earth than Jupiter. Some may turn out to be circling in very far-flung orbits around distant stars. All would be dim and supremely cold. But, like the ocean moons in our own solar system, a few might still hold the possibility of liquid water and life, if covered by a hydrogen-rich atmosphere. Or the planets might themselves have watery moons.

The conditions would need to be just right—a large moon orbiting an even larger planet, a thick atmosphere, and cosmic rays bombarding that atmosphere to trigger chemical reactions. If the parent planet's gravity tugs on the exomoon, it could warm it up the way Jupiter's gravity heats up its moon Io.

Astronomers worldwide are eagerly awaiting the launch of the Nancy Grace Roman Space Telescope later in the 2020s. With a field of view far greater than Hubble, it will perform microlensing surveys to spot exoplanets near and far—including the lonely worlds between stars.

—PATRICIA S. DANIELS

You have to be very good on a rope, and you have to be a pretty sturdy individual. And then you have to be a diver," Macalady told me. When she first visited Frasassi, she had zero caving experience; now she's a skilled caver and speaks fluent Italian.

In the aquifer's depths, a layer of clear fresh water sits atop salt water that's saturated with toxic hydrogen sulfide. That lower layer bubbles up from deep within the Earth, and in addition to being smelly, it's anoxic—there's no dissolved oxygen to power microbial metabolisms. "In that hydrogen sulfide layer, we expected to find almost nothing," Macalady said. "Instead, we found a forest of microbes unexpectedly making a living."

In 2004, Italian divers discovered that the toxic layer in one of the Frasassi lakes, Lago Infinito, was surprisingly inhabited. Black, stringy biofilms—cooperative communities of microbes—hung from the underwater rocky ceiling like tattered Gothic curtains, some of them three feet long. In the lake's seemingly sterile depths, the divers were spooked and fled. "If you see alien cave goo, your first instinct is to turn around," Macalady said, deadpan.

LEARNING

THE LIMITS OF LIFE

The Frasassi microbes Macalady was curious about are autotrophs, or organisms that make their own food from gases and minerals. Preliminary work points to thousands of species collaborating to craft these eerie appendages. Some have recognizable genetic sequences, but none are fully known to science. And a good fraction are what Macalady called "genetic dark matter"—the single-celled unknown. Finding those unknowns is not unprecedented, but as scientists sequence more microbes, it's "more rare to find a community that has so much of this genetic dark matter," she said.

On this trip, Dani Buchheister, one of Macalady's graduate students, was hoping that the divers could collect biofilms so she could try to coax the mysterious autotrophs to grow in a lab. If she could do that, perhaps she could figure out what they are and solve the mystery of how they survive in these pungent, suffocating waters—with an eye to what that means for life farther afield. "[I've become] more and more convinced that if we find life in our solar system, it's probably in subsurface environments," said Buchheister, who grew up in Florida.

And that's how we found ourselves on the muddy edge of Lago dell'Orsa, an 80-foot rappel from the tourist walkway above. Two divers had sunk beneath the cold, aquamarine waters to collect biofilm samples, but 10 minutes later, only one diver, Kenny Broad, had returned. He went back to search for his partner, Nadir Quarta, among the poorly mapped passages.

After an eternity (or perhaps several minutes), we heard the gurgling of two divers returning to the surface. "Well, that was slightly terrifying," Broad said to the crew as he floated. The 2011 Rolex National Geographic Explorer of the Year and an environmental anthropologist at the University of Miami, Broad was one of a handful of divers qualified for this expedition, which required proficiency with a rebreather as well as experience with underwater photography and scientific sample collection. Quarta had driven down from Switzerland. As they dived, the pair got split up, with Quarta heading down a passage he called "a mess." Underwater, amid poor visibility, Broad couldn't find him—a worrying situation during any dive. Once it was clear they were okay, Macalady asked, "Did you see some fuzzy gray stuff?"

During Broad's second dive in frigid Lago dell'Orsa—a 65-foot plunge that he described as "pretty shallow"—he filled five large syringes with diaphanous biofilms, material so fragile it fell apart at the slightest jiggle. Back at the field station, he thawed and stretched. "It wasn't visually stunning," Broad reflected. "But there's this otherworldly element to caves that's undeniable. They're a portal to some other place."

When we met up with the others in the field station's lab, Macalady had some of the biofilms under a microscope. Dark clumps of pyrite punctuated their cottonlike, wispy gray tendrils—perhaps the telltale sign of a primordial metabolic pathway that converts smelly hydrogen sulfide into pyrite and hydrogen gas. Scientists suspect such an energetically cheap pathway powered the first life-forms on Earth, but no one has found any evidence for it. "Those are pyrite framboids, the waste product of this metabolism we're looking for," Macalady said, indicating the dark sphere. "It's a clue."

Solving that mystery and identifying the microbes will depend on whether Buchheister can grow them in a lab. That might be tricky for multiple reasons, one being that the team's preliminary estimates suggest these microbes divide on very long timescales, perhaps once every century or millennium. There's so little energy available in the depths of the

Using a remotely operated vehicle, scientists collect vent fluid from an active black smoker two and a half miles beneath the Arctic Ocean.

Frasassi aquifer that, like the glass sponges populating the Arctic abyss, these microbes have figured out how to survive on what they can get.

"That's kind of my sweet spot," Buchheister said. "Learning about the limits of life on Earth, the weird places they live, and how that information might influence where and how we search beyond Earth."

LIFE IN THE ICE

When I returned to Longyearbyen in March 2023, Kalenitchenko—who'd been aboard the *Kronprins Haakon*—was collaborating with Kevin Hand, a National Geographic Explorer and astrobiologist at JPL. A lot had changed since we last met. And Ramirez-Llodra, co-chief scientist from my cruise, and her crew had revisited the Aurora vent field in 2021 and collected the deep-sea lifeblood erupting from two black smokers. Those vents now had names: Enceladus and Ganymede, the latter a reference to Jupiter's icy moon, the largest in the solar system.

But our target on this trip was terrestrial ice—in the pingos—not sea ice. Over several field seasons, this team will collect ice cores and characterize the microbial communities populating three sites near town—both in the pingos' reservoirs and locked in their ice. The work already suggests these microbes share characteristics with species that normally live in deep-sea biodiversity hot spots. That's because what occurs in the pingos is "what we see in the deep sea," Kalenitchenko said.

As in the Frasassi aquifer, hydrogen sulfide leaves its smelly calling card in some pingo reservoirs. The ice is full of methane, a compound Hand referred to as "life's first lunch." These two molecules can be both microbial food and microbial waste. Appropriately, the metabolic pathways that consume and excrete hydrogen sulfide and methane dominated the chemistry of the earthly environments I visited for this story. These pathways are arguably among the simplest and least energetically demanding in the single-celled repertoire, meaning they are, perhaps, a good model for what might arise elsewhere.

By melting ice cores and studying that methane, Hand can determine whether the gas is biologically produced or is the work of geology and chemistry—the exact kind of

Chance of Rain on Titan: 100% METHANE

As NASA's ingenious Dragonfly rotorcraft buzzes over the surface of Titan when it arrives in 2034, it might experience one of the solar system's rarest events: liquid rain. The cloudy moon is the only known world aside from Earth that experiences this kind of precipitation. It's not H_2O but CH_4, or methane, an odorless hydrocarbon that takes liquid form only in extremely cold temperatures. The Cassini spacecraft spotted fresh liquid methane on Titan's surface in 2016, implying a recent (and rare) methane shower. This kind of rain may descend only once every Titan year, equivalent to 29 Earth years, so Dragonfly may or may not arrive in a rainy season. Given the hazy sun through Titan's methane clouds, the rain might even bring broad rainbows—though they would be visible only through infrared goggles.

—PATRICIA S. DANIELS

A NASA technician studies the vault, or "brains," of the Europa Clipper spacecraft at California's Jet Propulsion Laboratory (JPL).

science we might someday do with a slice of alien ice. But identifying life's work on another world isn't as simple as studying the carbon atoms in methane. Remotely finding the fingerprints of extraterrestrial biology, perhaps locked in ice or floating in a teaspoon of seawater, will mean looking for multiple kinds of evidence of life.

On moons like Europa and Enceladus with chemistries like Earth's, scientists can make educated guesses about what they're looking for. On Titan's surface, which Dragonfly will explore, the terrain might look deceptively Earthlike, but the chemistry is anything but familiar. There it's so cold that liquid methane and ethane fill the lakes and seas, while the landscape is rock-hard water ice. Identifying life's fingerprints there will be a different challenge. "One just has to be very patient in the outer solar system," said planetary scientist Elizabeth "Zibi" Turtle of Johns Hopkins University Applied Physics Laboratory, who's spearheading the Dragonfly mission to Titan. "It's the nature of the part of the solar system we're exploring."

Learning whether these moons—or any world in the solar system—are inhabited is worth the challenge. "We need to find out," said JPL's Robert Pappalardo, science lead for the Europa Clipper mission. "It really could bring a new Copernican revolution in our understanding of our place in the universe."

TITAN

Like Europa and Enceladus, Saturn's moon Titan is believed to have an interior saltwater ocean. But that's not the only reason the solar system's second largest moon may harbor life. Where Earth has a water cycle, Titan's methane cycle rains complex molecules onto its surface, creating methane lakes, rivers, and massive surface dunes made not of sand but of granular hydrocarbons.

DRAGONFLY

Titan's atmosphere is 60 percent denser than Earth's at its surface, making the moon a perfect candidate for drone exploration. Powered by a 94-pound radioactive fuel source, NASA's thousand-pound Dragonfly octocopter, scheduled to arrive in 2034, will take images and analyze surface samples across Titan's exotic terrain.

Onboard lab
Thick insulation and a durable power source will warm Dragonfly's instruments to between -40°F and -22°F so they can operate on Titan's -290°F surface. A mass spectrometer will analyze surface samples for signs of life.

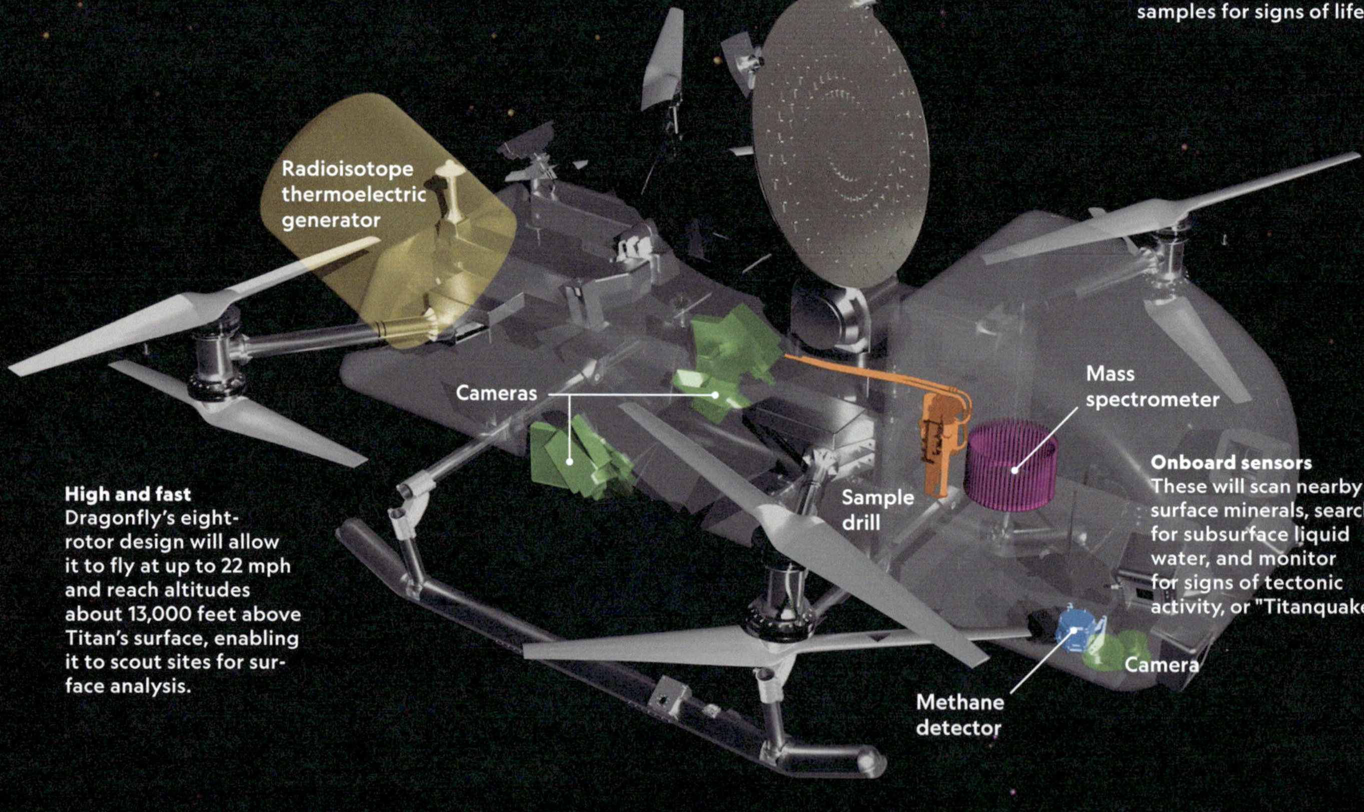

High and fast
Dragonfly's eight-rotor design will allow it to fly at up to 22 mph and reach altitudes about 13,000 feet above Titan's surface, enabling it to scout sites for surface analysis.

Onboard sensors
These will scan nearby surface minerals, search for subsurface liquid water, and monitor for signs of tectonic activity, or "Titanquakes."

JASON TREAT, NGM STAFF; SEAN MCNAUGHTON
ART: DANA BERRY. SOURCES: STEVEN D. VANCE, NASA/JPL; STEVE GRIBBEN, JOHNS HOPKINS UNIVERSITY APPLIED PHYSICS LABORATORY; NASA

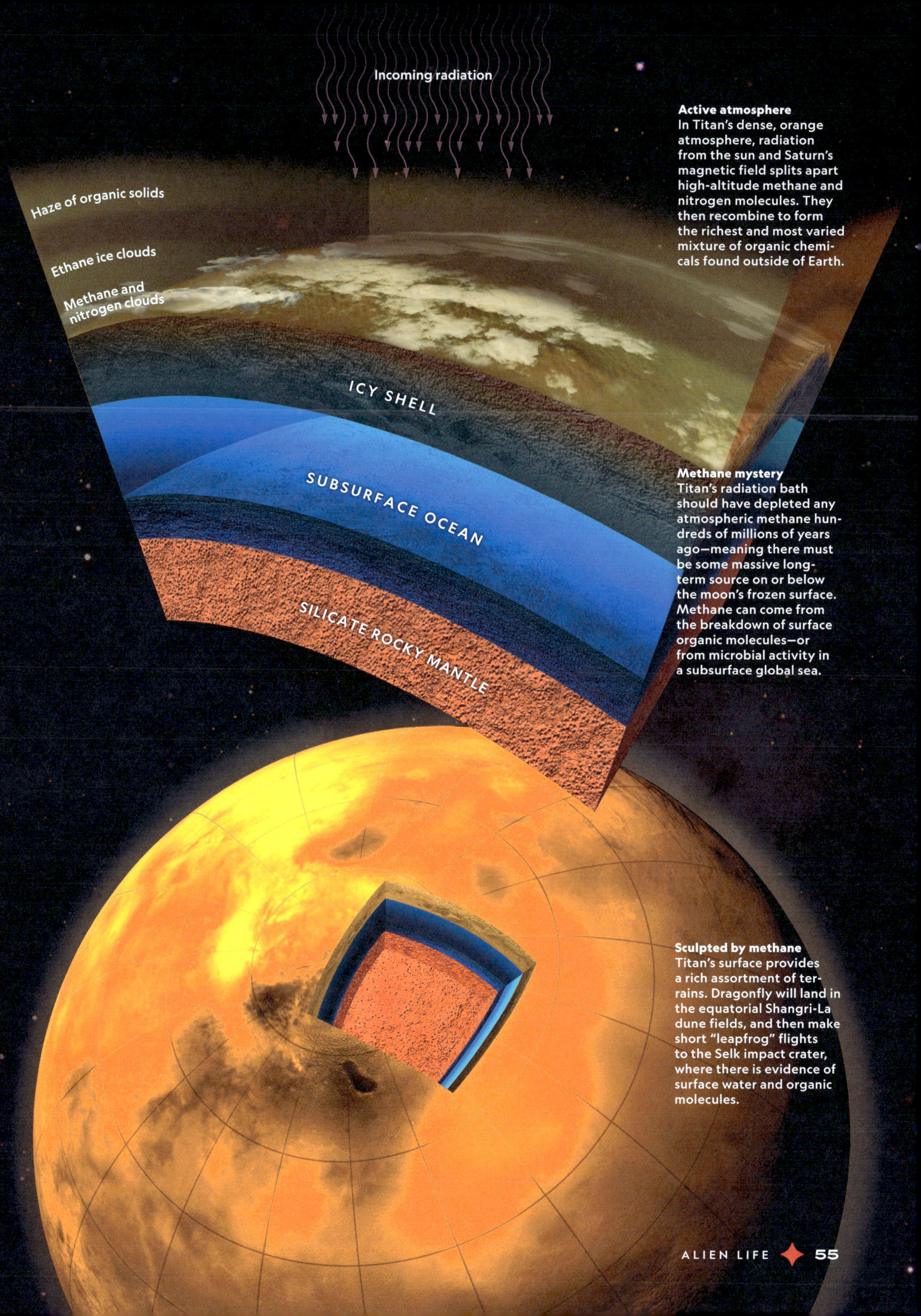

Active atmosphere
In Titan's dense, orange atmosphere, radiation from the sun and Saturn's magnetic field splits apart high-altitude methane and nitrogen molecules. They then recombine to form the richest and most varied mixture of organic chemicals found outside of Earth.

Methane mystery
Titan's radiation bath should have depleted any atmospheric methane hundreds of millions of years ago—meaning there must be some massive long-term source on or below the moon's frozen surface. Methane can come from the breakdown of surface organic molecules—or from microbial activity in a subsurface global sea.

Sculpted by methane
Titan's surface provides a rich assortment of terrains. Dragonfly will land in the equatorial Shangri-La dune fields, and then make short "leapfrog" flights to the Selk impact crater, where there is evidence of surface water and organic molecules.

Tidal forces
The orbits of three of Jupiter's four inner moons regularly align, amplifying tidal forces from Jupiter's immense gravity. Europa is pulled and stretched by both Jupiter's gravity and that of these larger moons.

EUROPA

Scientists suspect that hiding beneath Europa's frozen surface is a saltwater ocean with twice the volume of all Earth's oceans combined. Flying as close as 16 miles above the moon's surface, NASA's Europa Clipper spacecraft, scheduled to launch in late 2024, will use ice-penetrating radar and other sensors to seek out signs of liquid water within Europa's ice shell and study organic molecules on its surface.

JASON TREAT, NGM STAFF; SEAN MCNAUGHTON
ART: DANA BERRY. SOURCES: STEVEN D. VANCE, NASA/JPL; NASA

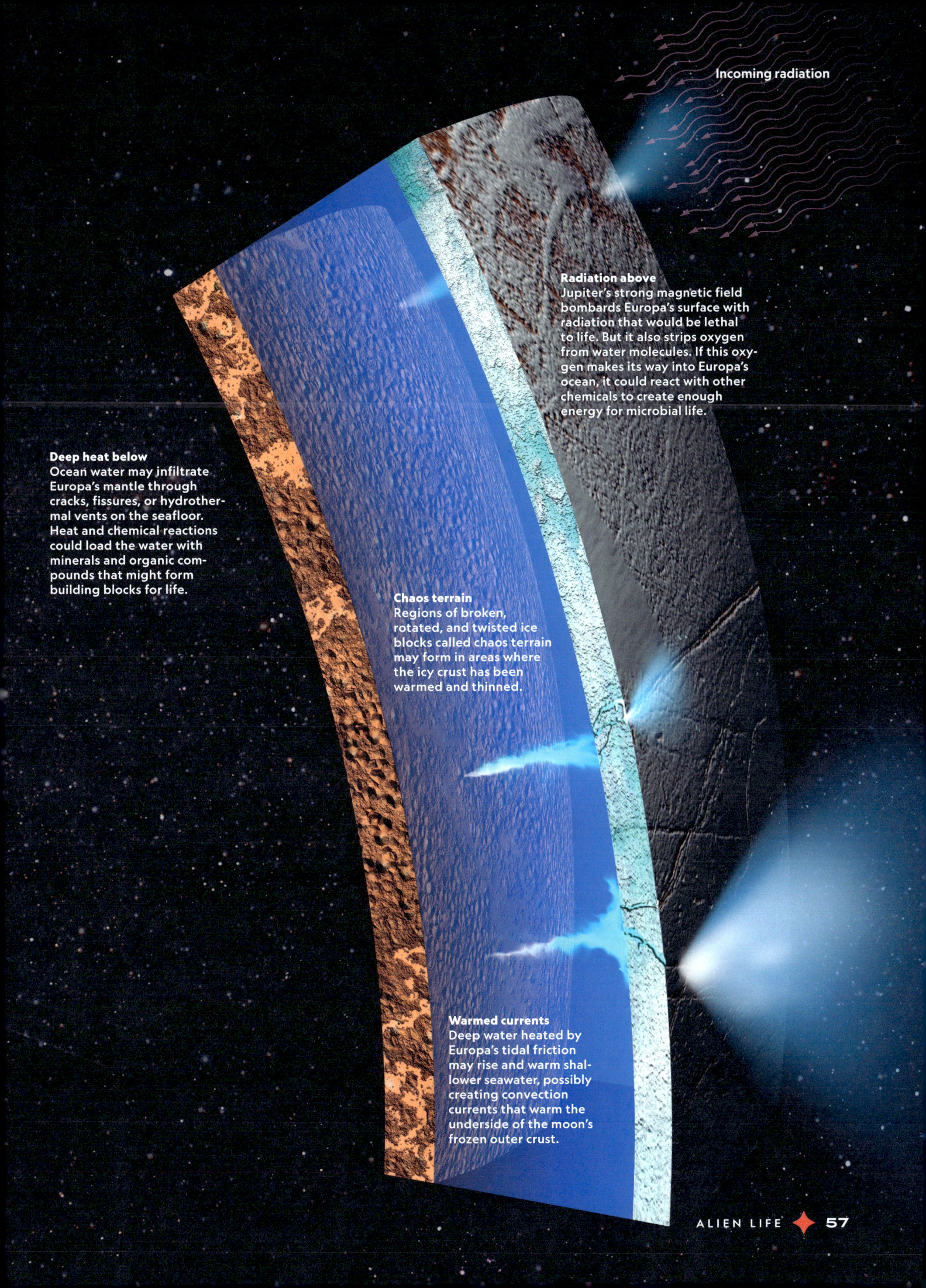
Incoming radiation
Radiation above
Jupiter's strong magnetic field bombards Europa's surface with radiation that would be lethal to life. But it also strips oxygen from water molecules. If this oxygen makes its way into Europa's ocean, it could react with other chemicals to create enough energy for microbial life.
Deep heat below
Ocean water may infiltrate Europa's mantle through cracks, fissures, or hydrothermal vents on the seafloor. Heat and chemical reactions could load the water with minerals and organic compounds that might form building blocks for life.
Chaos terrain
Regions of broken, rotated, and twisted ice blocks called chaos terrain may form in areas where the icy crust has been warmed and thinned.
Warmed currents
Deep water heated by Europa's tidal friction may rise and warm shallower seawater, possibly creating convection currents that warm the underside of the moon's frozen outer crust.

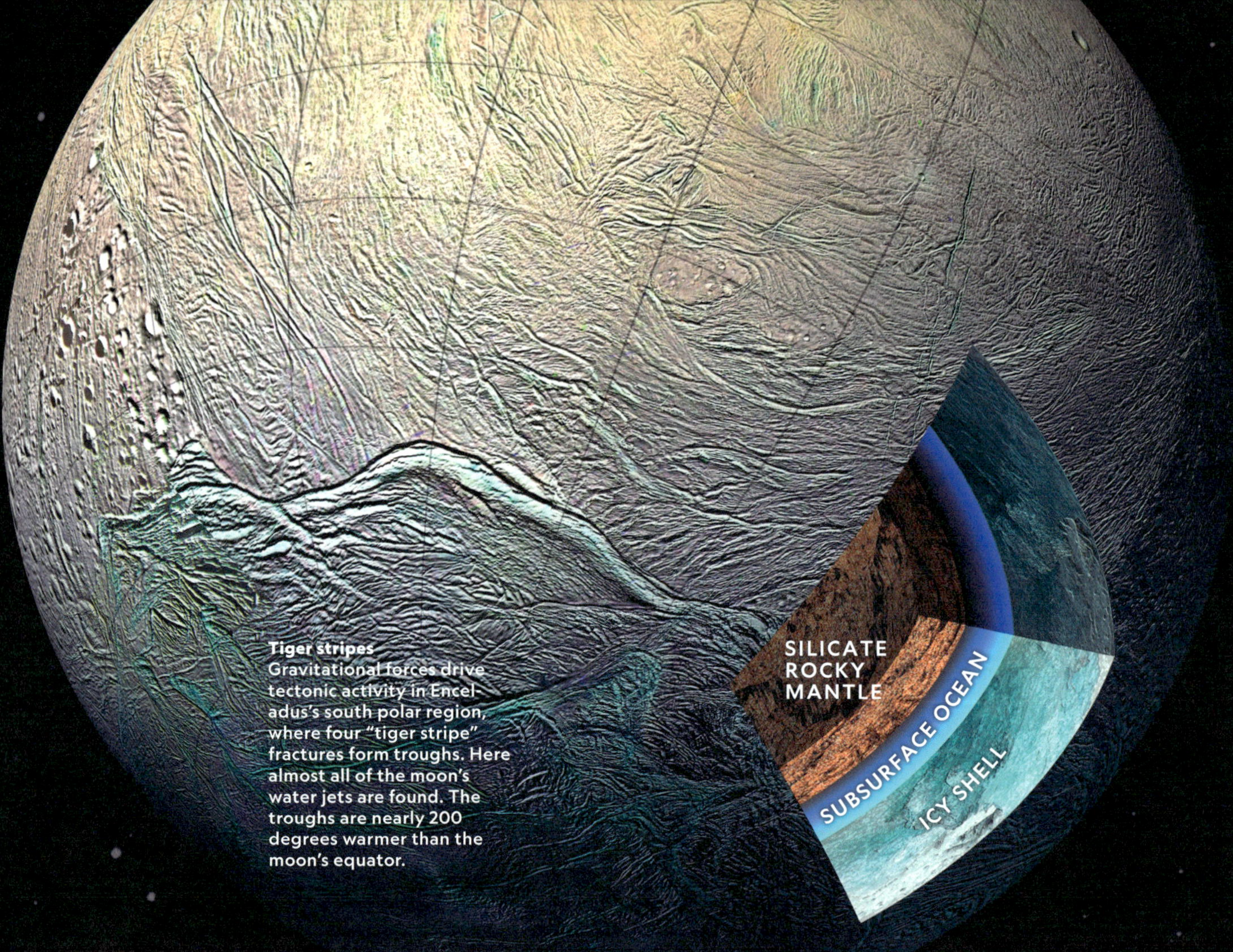

Tiger stripes
Gravitational forces drive tectonic activity in Enceladus's south polar region, where four "tiger stripe" fractures form troughs. Here almost all of the moon's water jets are found. The troughs are nearly 200 degrees warmer than the moon's equator.

ENCELADUS

Scientific observations strongly suggest that Titan, Europa, and other moons in the solar system have liquid water beneath their surfaces. Saturn's moon Enceladus, however, showcases its ocean by spraying so much salty water into space that it forms the planet's diffuse, massive, and beautiful E ring.

Feeding the E ring
Enceladus orbits in resonance with larger moon Dione, which amplifies tidal forces on the tiny moon created by Saturn's gravitational field. This interaction stretches and compresses Enceladus, generating heat that fuels chemical reactions and drives its high-powered surface jets.

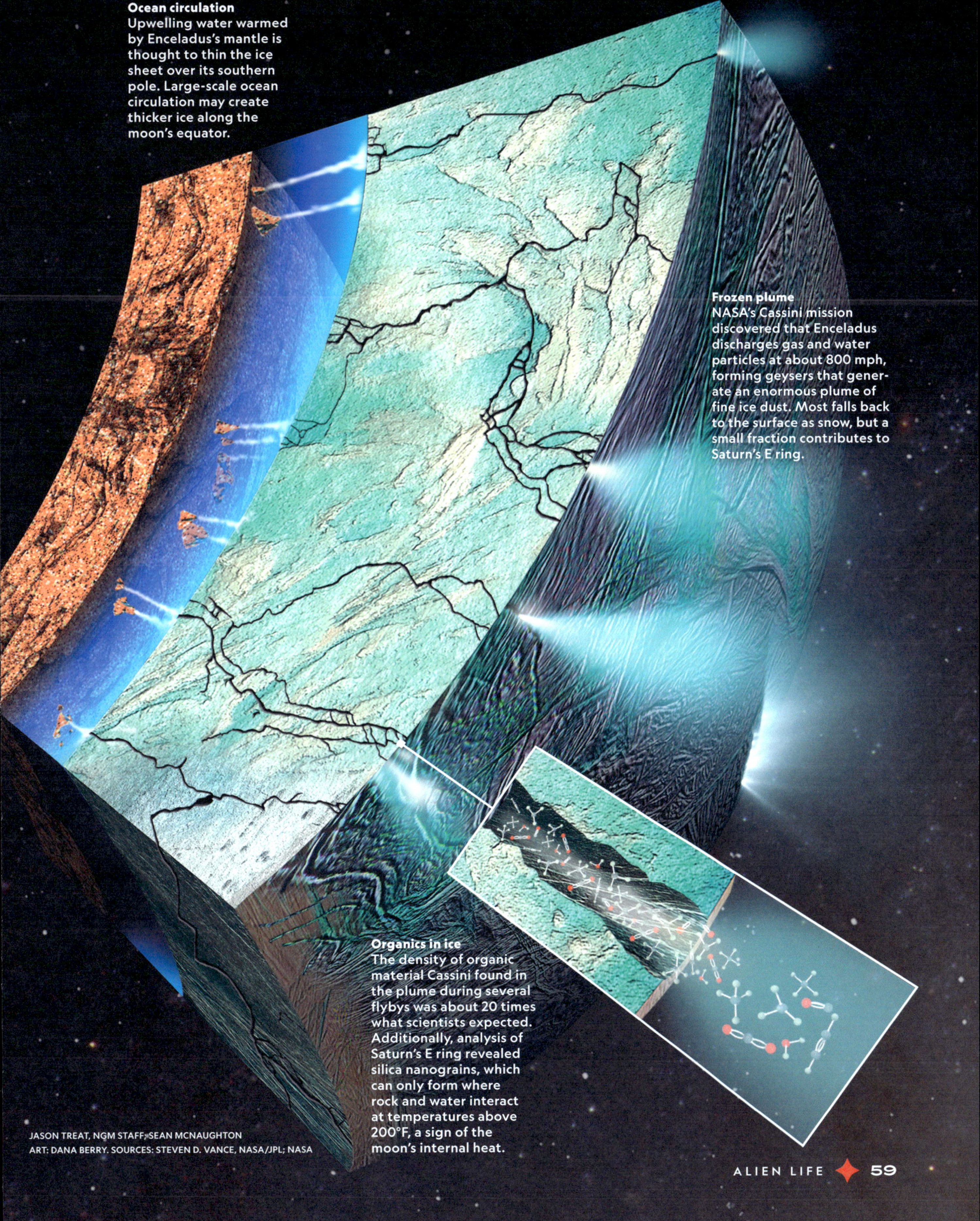

Ocean circulation
Upwelling water warmed by Enceladus's mantle is thought to thin the ice sheet over its southern pole. Large-scale ocean circulation may create thicker ice along the moon's equator.

Frozen plume
NASA's Cassini mission discovered that Enceladus discharges gas and water particles at about 800 mph, forming geysers that generate an enormous plume of fine ice dust. Most falls back to the surface as snow, but a small fraction contributes to Saturn's E ring.

Organics in ice
The density of organic material Cassini found in the plume during several flybys was about 20 times what scientists expected. Additionally, analysis of Saturn's E ring revealed silica nanograins, which can only form where rock and water interact at temperatures above 200°F, a sign of the moon's internal heat.

JASON TREAT, NGM STAFF; SEAN MCNAUGHTON
ART: DANA BERRY. SOURCES: STEVEN D. VANCE, NASA/JPL; NASA

IS THERE LIFE

RECENT FINDINGS FROM NASA ROVERS LOOK TANTALIZINGLY LIKE ALIEN LIFE SIGNATURES—BUT MARS HAS FOOLED US BEFORE, AND SCIENTISTS HAVE YET TO FULLY EXPLAIN ALL ITS PLANETARY WORKINGS.

WORDS BY **NADIA DRAKE**

ON MARS?

NASA's Perseverance rover looks down at a sample tube filled with Martian rock. A future mission will collect it.

FOR SCIENTISTS SEARCHING for alien life-forms, the siren song of Mars is climbing toward a crescendo. Multiple observations made by rovers on the red planet could bear the signatures of microbes—a possible indication that Earth is not the only refuge for life in the solar system.

In January 2022, NASA's Curiosity rover observed a mixture of carbon isotopes in the rocks of Gale crater that, if seen on Earth, would be a sign of life. The rover has also witnessed both random and seasonal surges of methane, a gas on Earth that is predominantly produced biologically.

About 2,300 miles away in Jezero crater, NASA's Perseverance rover spied strange purple coatings on the crater floor's rocks. These coatings are widespread and resemble desert varnishes on Earth that grow in the presence of microbes.

For now, though, scientists aren't ready to conclude that our vermilion neighbor was once inhabited. Just about every alluring hint of biology could also be explained by some as yet unfamiliar aspect of Mars's geology or chemistry—there's just so much we don't know about how the planet works, and how nonliving phenomena could be masquerading as life's fingerprints.

"This is an alien world that we're looking at, and so who knows what we haven't even thought of," says Curiosity deputy project scientist Abigail Fraeman of NASA's JPL.

IS THERE LIFE ON MARS?

A NEW ERA OF MARS EXPLORATION began in 2012 when Curiosity landed in Gale crater, where it almost immediately identified a swath of rocks that can form only when water is present.

Over the years, Curiosity has uncovered evidence in the crater for numerous organic molecules—the chemical building blocks for carbon-based life-forms. And it has spotted signs of ancient hydrothermal activity, where heat and chemical compounds mixed with flowing water, creating possible energy sources.

The rover has also determined that methane gas in the crater rises and falls as the seasons change. Such a fluctuation on Earth would be a strong sign of beings with active metabolisms.

However, none of these observations have so far been linked to biology, and there's always a chance that processes we don't fully understand are mimicking the signatures of life.

"Most carbon-related processes on Earth's surface are biological, so to try and change our mindset around and think about a world where that might not be true is really a challenge," says astrobiologist Christopher House of Penn State University.

> "THIS IS AN ALIEN WORLD THAT WE'RE LOOKING AT, AND SO WHO KNOWS WHAT WE HAVEN'T EVEN THOUGHT OF."
>
> ABIGAIL FRAEMAN, NASA JPL

THE CURIOUS CASE OF MARTIAN CARBON

CURIOSITY'S WEIRDEST, most tantalizing observation has been this: In multiple rock samples from various locations in the crater, the rover found organic compounds containing odd ratios of carbon isotopes, or atoms of the same element that contain different numbers of neutrons in their nuclei.

On Earth, organisms use the lighter form of carbon in metabolic or photosynthetic reactions, leading to more of the lighter form than the heavier.

And in five locations in Gale crater, scientists found that lighter carbon isotopes were much more abundant than their heavier cousins, relative to what scientists have seen in the Martian atmosphere and in meteorites. The observations resemble carbon ratios collected from Earth. "These really depleted carbon isotope results are so intriguing. So compelling. On Earth, the only way you do this is with biology," says astrobiologist Amy Williams of the University of Florida, though House, who led the analysis, says the story is far from clear.

A MYSTERIOUS COATING ON THE ROCKS

NASA'S PERSEVERANCE ROVER arrived at Jezero crater on Mars in 2021, and it's also on the hunt for signs of ancient life.

During its travels through Jezero, Perseverance spied numerous rocks with a purple, iron-rich coating. Purdue University's Bradley Garczynski, who is studying the coating, says it's unlike anything that rovers have spotted

A distant beacon, Mars, next to the Milky Way, glows over California's Sierra Nevada.

on Mars before—even though rocks with different coatings have been seen on other parts of the planet.

On Earth, such coatings are often observed in deserts, where conglomerates of rock-munching microbes thrive.

"They're really intriguing, and they're certainly on Earth of biological interest, so by translation they are then of great astrobiological interest to us when we see them forming on other worlds," Williams says.

Kennda Lynch, who studies terrestrial analogs of Martian environments at the Lunar and Planetary Institute, says finding biosignatures in the Jezero rock varnishes wouldn't be impossible. "Microbes do amazing things. They put coatings and varnishes on the rocks because they like to eat the rock," she says.

However, Lynch says, context is crucial for properly interpreting an observation. Even on our own planet, investigators need to rigorously evaluate whether such a material was produced by life or by some other process. That's a much tougher question to answer from afar.

"It's a wonderfully complicated and complex system that we're exploring on Mars," Fraeman says.

Scientists say the next step in probing Mars for life is bringing bits of the planet back to labs on Earth, where the sharpest instruments available can search for answers to one of humanity's oldest questions. The Perseverance rover is already busy collecting the first set of samples, which could contain evidence that microorganisms lived in Jezero crater billions of years ago.

The camera aboard Webb, which sees in the near-infrared light spectrum, captures the growth of a protostar 460 light-years from Earth.

PART 3

THE MYSTERIES OF DEEP SPACE

The scientific marvel that is the James Webb Space Telescope was expected to make significant discoveries about the early universe, and so it has. But Webb has also delivered another kind of news. The universe is gorgeous. Glowing gases, swirling spiral galaxies, filmy nebulae, and vast vistas of stars in different hues: Who knows what other surprises lie ahead?

WORDS BY
JAY BENNETT

TRAVELING TO ANCIENT GALAXIES

THE JAMES WEBB SPACE TELESCOPE, THE MOST POWERFUL EYES EVER PUT IN THE SKY, IS ALREADY TRANSFORMING OUR UNDERSTANDING OF THE COSMOS.

In one of the deepest views into the universe ever achieved, Webb reveals thousands of objects, including the bright, light-bending star cluster in the center.

Astronomers Marcia and George Rieke have played key roles in the development and operations of Webb instruments: the near-infrared camera, or NIRCam, and the mid-infrared instrument, or MIRI.

WHEN THE UNIVERSE WAS YOUNG,

MORE THAN 13 AND A HALF BILLION YEARS AGO, no stars shone in the abyss. It was a time when the cosmos was filled with hydrogen and helium gas, the raw material for all the worlds to come.

A mysterious substance known as dark matter existed too, its gravity pulling the gas into an elaborate web. As things expanded and cooled, some of the dark matter consolidated in immense orbs, driving the gas to their cores. The rising gravitational pressure within these halos, as astronomers named them, forced hydrogen atoms to fuse into helium, igniting the primordial universe's first stars.

Scientists have been filling in the universe's origin story for decades, but in the past few years, the largest and most advanced space telescope ever built has rewritten the first chapters. Ancient galaxies glimpsed by the James Webb Space Telescope (JWST) are brighter, more numerous, and more active than anticipated, revealing a frenetic opening to the saga of space and time.

Webb cannot see the first stars, though, as they weren't bright enough to detect individually. These early monsters blazed hot and grew immense before erupting in supernovae a few million years after flaring to life—a blip in astronomical time.

"It's just so crazy fast," said Tom Abel, a computational cosmologist projector at the Kavli Institute for Particle Astrophysics and Cosmology at Stanford University.

The supernovae transformed the universe, generating new elements—oxygen to make water, silicon to build planets, phosphorus to power cells—and scattered throughout the expanse. The first stars also broke apart the atoms of the surrounding hydrogen

Ultraviolet radiation drives chemical reactions in the Orion Nebula, where Webb found the carbon compound methyl cation, one of life's building blocks.

Only four of the five galaxies known as Stephan's Quintet are truly close together. The detail from Webb lets scientists better study galactic evolution.

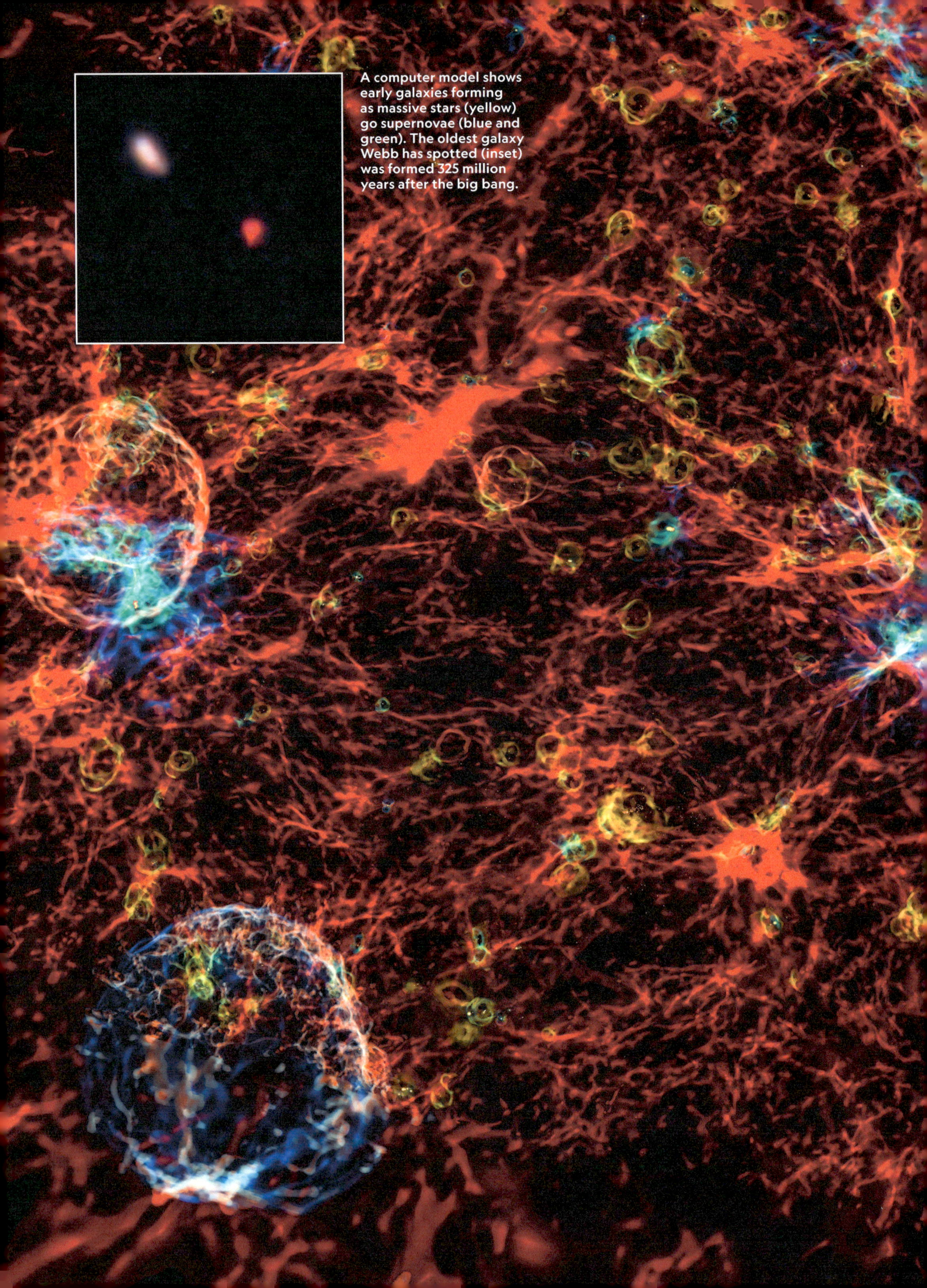

A computer model shows early galaxies forming as massive stars (yellow) go supernovae (blue and green). The oldest galaxy Webb has spotted (inset) was formed 325 million years after the big bang.

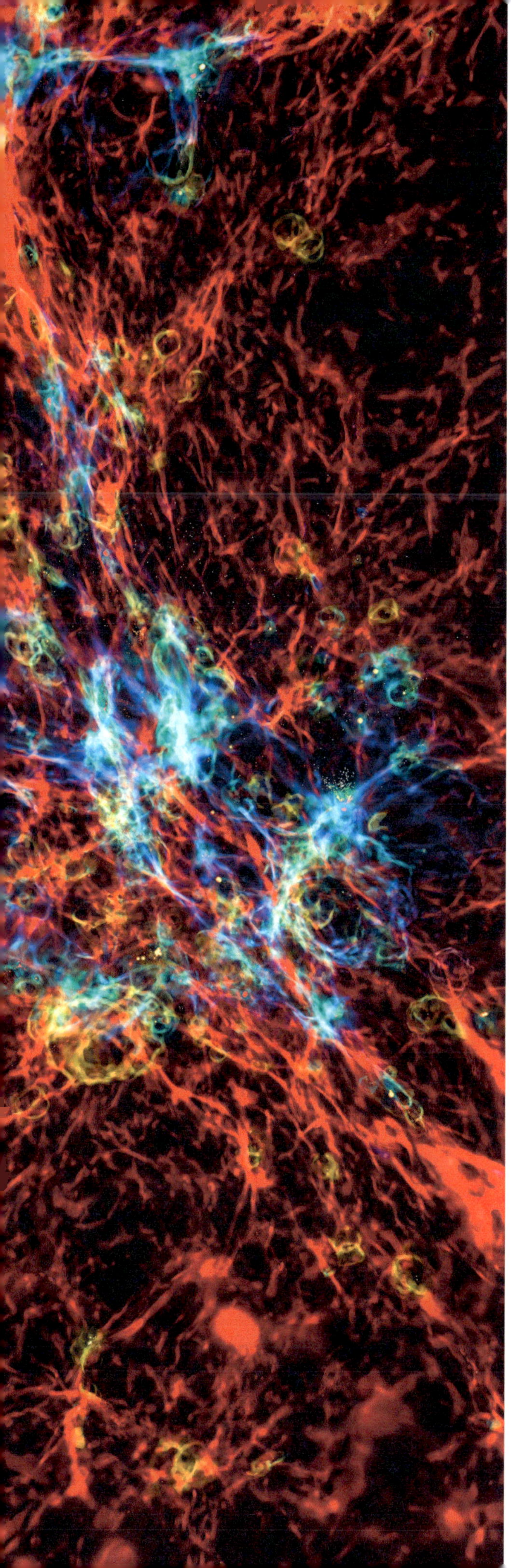

gas, burning away the cosmic haze and making things transparent—a key time known as reionization. As the fog lifted, pockets of stars merged, swirling into bigger and bigger assemblages, including the seed of our own Milky Way.

Abel and his colleagues, in the 1990s, started using computer simulations to help determine that the first astronomical residents had to have been stars. But eventually Abel, thinking there was nothing more to learn, moved on from star-birth simulations.

Then came Webb.

LOOKING

INTO THE PAST

Launched on Christmas morning in 2021, Webb is now positioned nearly a million miles from Earth. Its 21-foot-4-inch gold-coated primary mirror captures the light of ancient galaxies, which has been traveling through space for more than 13 billion years, revealing the galaxies as they were in the distant past.

Astronomers expected to find some of these infant galaxies with Webb. They didn't expect to find so many—or that the discoveries could shake their understanding of galactic history.

The deepest ever galaxy survey of the universe kicked off in September 2022, when an international collaboration known as JADES, or the Webb Advanced Deep Extragalactic Survey, began using Webb to observe patches of sky for dozens of hours at a time. Two weeks after observations began, some 50 astronomers gathered in Tucson at the University of Arizona to discuss the first results.

Packed into a classroom, the scientists, from tenured professors to twentysomething graduate students, were preoccupied with the mosaic on their laptops: hundreds of images freshly captured by Webb and stitched together. The picture, shared with the team only days before, contained tens of thousands of galaxies and other celestial objects. Excited murmuring ran through the group as they pointed out things to one another that had never been seen: active star-forming regions, glowing galactic centers where black holes might be, and reddish blobs of light from galaxies so distant only Webb could spot them.

"This is a little bit like kids in the candy shop," said Marcia Rieke, an astronomer at the university and one of the leaders of the collaboration.

Unlike the Hubble Space Telescope—our previous window into the universe's distant past—Webb was designed to observe in the infrared spectrum, which makes it ideal for capturing early starlight. Those rays left their source as ultraviolet but were stretched to redder wavelengths by the expansion of the universe, a phenomenon known as redshift. The higher the redshift, the farther and older the target.

Rieke managed the proceedings with a combination of delight and sage reflection. Besides being a lead scientist on JADES, she's the principal investigator of Webb's near-infrared camera, or NIRCam—the source of the mosaic of galaxies on everyone's laptop. She oversaw its design, a 330-pound assemblage of mirrors, lenses, and detectors to drink in the light of the universe and study it through different filters.

"These images are everything we could have hoped for," she said.

The team wasn't able to change where the telescope was pointing, so it had to target objects already in the field of view—and thanks to a bit of luck, four faraway galaxies detected by NIRCam were sitting in the right spot. Two of those, observations would later confirm, were more distant and ancient than any known before.

The most far-flung of the bunch, called JADES-GS-z13-0, had been formed only 325 million years after the big bang. "I still have the Slack message where I first saw this object in the data and sent it to the group," said Kevin Hainline, an astrophysicist at the University of Arizona. "In the craziness of it, I didn't realize the profundity of this moment of sitting there and being like, Oh, that's the farthest galaxy that humans have ever seen."

REVELING

IN THE CONFUSION

Two things are already clear about these early galaxies: There are more of them than expected, and they are surprisingly bright for their age. "However star formation gets going in the early universe, it's not quite like how we might have predicted," Rieke says.

VYING FOR ROOM AT THE EDGE OF SPACE

The many satellites circling Earth are increasing light pollution, making it harder to get clear data.

The new generation of satellites orbiting close to Earth allows for high-speed internet across the planet. But booming connectivity begets congestion: There are more than 11,000 satellites in the skies. Within this decade, that number could top 100,000. Appearing as streaks of light in telescopic photography, satellites can obscure stars and overexpose images, making it a challenge for astronomers to study the cosmos.

A recent study in *Nature Astronomy* found that, on average, some 3 percent of individual images captured by the Hubble Space Telescope between 2002 and 2021, including those at right, contained at least one satellite trail. While the rate is still relatively low, the study estimates that by the 2030s, up to 50 percent of Hubble's photographs may be marred by satellite streaks.

SATELLITE

In this photo of the cosmos taken through a telescope, part of a satellite intrudes on the view. Satellite pollution is only expected to worsen.

FROM PIGEONS TO THE BIG BANG

Surprisingly mundane beings played a scientific role in a key astronomical discovery in the '60s: two pigeons. In 1964 Robert Wilson and Arno Penzias were getting an irritating hiss from their horn-shaped radio telescope in Holmdel, New Jersey.

Was it picking up radiation from nearby New York City? Or could the pigeons nesting in the telescope and decorating its surface with droppings be the source?

The astronomers cleared away the birds and their guano, but the hiss remained. They realized it had to be coming from space itself. In time, they identified the noise as microwave radiation from all quadrants of the universe, leftover heat from the big bang—a discovery that won them the 1978 Nobel Prize in Physics.

The Holmdel Horn Antenna, under threat by local development, has recently been protected and preserved within a 35-acre park. —PATRICIA S. DANIELS

One early galaxy, GN-z11 from some 440 million years after the big bang, is bright enough that Hubble spotted it in 2016. Now Webb has observed the object as well, including taking its spectrum with an instrument called NIRSpec.

"This one has everyone sort of confused and excited," says Emma Curtis-Lake, an astrophysicist at the University of Hertfordshire in England and a member of the NIRSpec team.

Certain elements create bright emission lines in a galaxy's spectrum, like fingerprints. The spectrum of GN-z11 revealed a surprising amount of nitrogen—confounding scientists, who can't explain its source.

GN-z11 may also host a supermassive black hole, which would be remarkable for this early time. It'd be "the most distant black hole that we've seen," Curtis-Lake says.

Obscured at the center of the bright galaxy, it was exposed by spectral lines that Curtis-Lake calls "little hidden monsters." These lines suggest that material is moving rapidly in a dense area, swirling at roughly a million miles an hour—the kind of thing you would expect to see near a black hole. But how one of these objects could have grown so rapidly remains unsolved.

TELESCOPE PIONEERS

"This ain't like it used to be," said Rieke's husband, George, as he stepped into a control room that doubled as a kitchenette. "No," Marcia agreed. "There's five times as many monitors."

The couple had offered to show me an old telescope in the mountains near Tucson, where they spent much of their early careers. Both astronomers at the University of Arizona, they met in 1972. The 61-inch telescope on Mount Bigelow was fairly new then, used to make maps of the lunar surface. It became one of the leading observatories in the budding field of infrared astronomy, a grandfather of sorts to Webb.

The Riekes helped facilitate this succession. While Marcia oversaw the development of NIRCam, George is the lead scientist on Webb's mid-infrared instrument, or MIRI. The Riekes would stay awake through the night, slowly adjusting the telescope manually to keep a target in sight as Earth rotated. Today most of the work can be done from laptops.

A diagonally split view of spiral galaxy NGC 628 shows it in Webb's infrared spectrum (top left) and Hubble's visible spectrum (bottom right).

In the 1970s Marcia and George used the telescope on Mount Bigelow to make some of the first infrared observations of the Milky Way's center. Scientists had assumed that part of our galaxy was "a collection of old, uninteresting stars," Marcia explained. But in infrared light, turbulent pockets of gas with rapid star formation were revealed. "That whole picture got changed," George added.

At the time, the infrared light of the cosmos was only just coming into view. New sensors tuned to it revealed this previously hidden part of the electromagnetic spectrum, which is the full range of light, from gamma rays to radio waves. The telescope on Mount Bigelow helped fill a gap in observations of the local universe, and Webb has similarly plugged a hole in our view of the deep cosmos.

But to fully understand our cosmic origins, we will need more than just Webb.

SEARCHING FOR THE EARLY GLOW

On a recent spring morning, I squinted in the sunlight on an expansive plateau between snowcapped volcanoes in Chile's Atacama Desert. Plastic tubes tickled my nostrils with the flow of oxygen, a requirement for anyone visiting the 16,400-foot-high site of the Atacama Large Millimeter/submillimeter Array (ALMA).

The sky was a deeper shade of blue up there, with fewer molecules in the atmosphere to scatter the light—the very thing that makes this place perfect for astronomy. Towering before me were dozens of four-story-tall radio

TO FULLY UNDERSTAND OUR COSMIC ORIGINS, WE WILL NEED MORE THAN JUST WEBB.

dishes, white sentinels scattered across the Chajnantor plateau. They pivoted in unison to lock onto a new target.

Among the most advanced radio observatories on the planet, ALMA is also one of the few tools capable of examining the early galaxies being discovered by Webb, albeit in a different light. Webb captures starlight punching through the dust of these galaxies, while ALMA searches for the glow of the dust itself, heated by the stars within.

"These first dust grains come from supernova explosions, so you can indirectly obtain information about the first supernova explosions and the first population of stars," said María Emilia De Rossi, an astrophysicist at the Institute for Astronomy and Space Physics (IAFE) in Buenos Aires.

ALMA has trained its radio dishes on some of the early galaxies, but in most of its first attempts, the array wasn't able to find any dust emissions. This could mean that the galaxies are in their infant stages and have not yet produced much dust through stellar explosions, or it could mean that some are actually closer than thought.

ALMA's first attempts to detect the galaxies discovered by Webb were only glances, short-duration observations slotted into its busy schedule. Astronomers plan to point the array at some of these galaxies for longer periods, searching for faint signals that could reveal how much dust they have generated and, crucially, how many heavy elements they have produced—an indication of how far along they are in galactic evolution.

Toward the end of my visit, I stopped by an enormous hangar at ALMA's operations facility, lower at 10,000 feet. Two of the towering radio dishes had been brought down from the high site on a 28-wheel transporter vehicle. Workers on lifts were busy replacing some of the dishes' components, part of a series of upgrades to make the observatory even more capable.

Soon the dishes would be returned to the plateau—ready to swing their gaze back to the firmament, primed to tackle the mysteries of primordial galaxies.

Newborn stars burst from cocoons of gas and dust in this Webb image of the Rho Ophiuchi cloud complex, the star-forming region closest to Earth.

THE EARLIEST BARRED SPIRAL GALAXIES

In this Hubble Space Telescope image, a reddish bar running through barred spiral galaxy NGC 1300 (at left) funnels gas to the center, triggering rapid star formation. Webb has found much older ones, below.

All ages below in years after the big bang:

2.8 billion | 3.1 billion | 4.2 billion

5.1 billion | 5.2 billion | 5.4 billion

THE MAKING OF A MASTERPIECE

Thousands of people from NASA, the European Space Agency, and the Canadian Space Agency worked for 17 years, spending billions of dollars, to create the James Webb Space Telescope. At the end it's safe to say all of them were in an agony of suspense as the immensely complicated instrument was launched from a spaceport in Kourou, French Guiana, in 2021.

Planning for Webb (originally called the Next Generation Space Telescope) began in the 1990s, even as the Hubble optical telescope was wowing the public with spectacular images of deep space. Astronomers knew they could learn even more about the universe if they could see it in infrared light, which carries information about the earliest stars, as well as about the atmospheres of distant planets.

Construction of the components—the 21.3-foot primary mirror, the tennis-court-size folding sun shield that protects it, the camera, and other scientific instruments—began in 2004. Because the telescope would operate in intense cold, the 2010s models of the delicate structures were tested in a cryogenic chamber at the Johnson Space Center in Houston. And because the whole contraption had to survive the stresses of launch and deployment, the pieces were also mounted on a framework, then shaken and blasted with sound waves.

On December 25, 2021, Webb launched atop an Ariane 5 rocket. Thus began what one scientist called "three weeks of terror" as the team waited to see if each delicate piece would deploy correctly. Within an hour, the telescope separated from the rest of the spacecraft and unfurled its solar array. Within a week, the hair-thin membranes of the sun shield unfolded. After two weeks, the gold-foil mirrors slowly swung into place. On January 24, 2022, Webb coasted into its orbit around the sun; its team breathed a sigh of relief and began to scan the infrared universe. —PATRICIA S. DANIELS

JULY 10, 2014
Engineers at Northrop Grumman in Redondo Beach, California, prepare to unfurl a test version of the five-layer sun shield, designed to keep the new observatory cool.

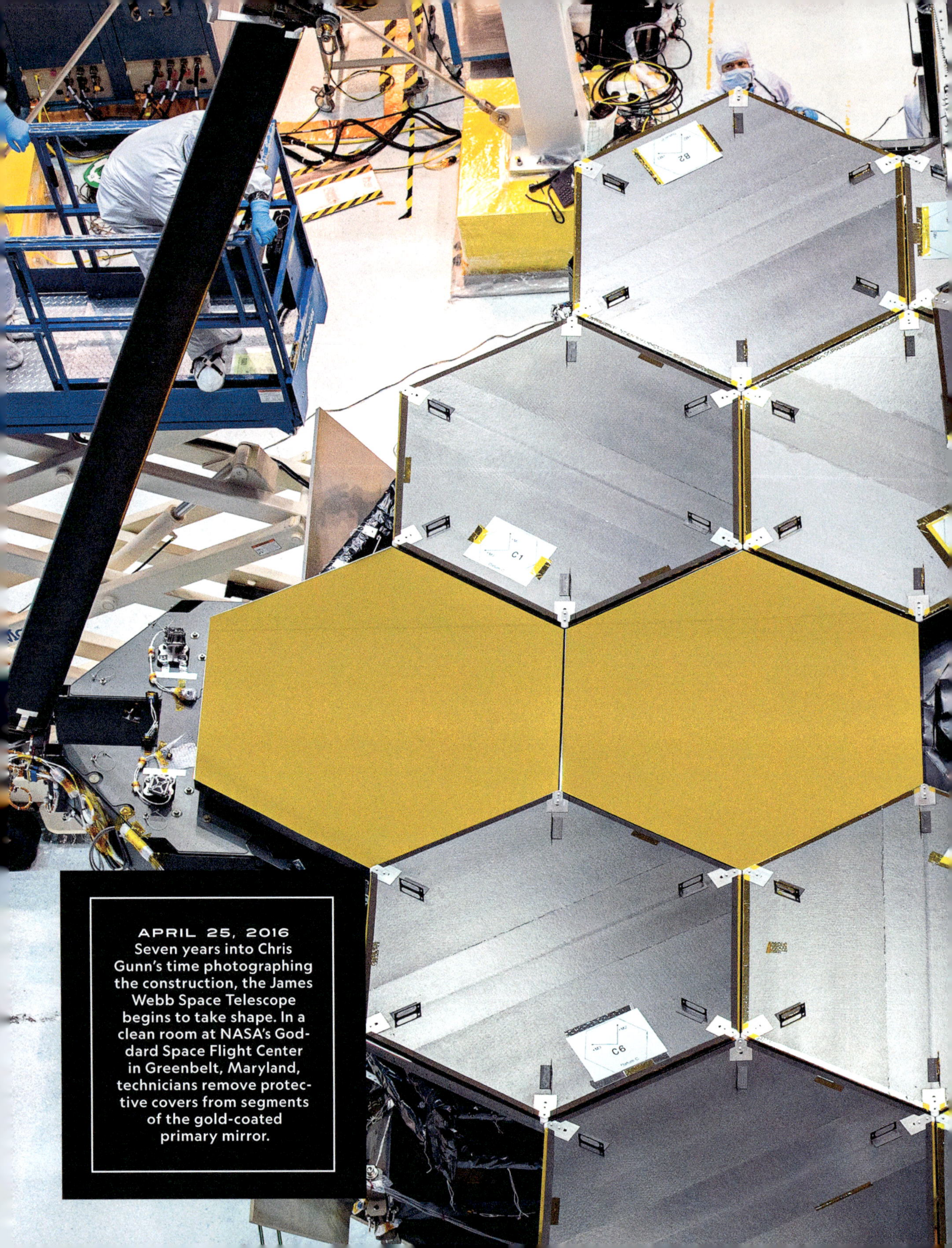

APRIL 25, 2016
Seven years into Chris Gunn's time photographing the construction, the James Webb Space Telescope begins to take shape. In a clean room at NASA's Goddard Space Flight Center in Greenbelt, Maryland, technicians remove protective covers from segments of the gold-coated primary mirror.

DECEMBER 1, 2017
From a lift eight stories high, Gunn captures Webb as it is moved out of the thermal vacuum chamber where it had undergone cryogenic testing for about a hundred days at NASA's Johnson Space Center in Houston. After the successful test, the excitement was palpable, Gunn says: "That much closer to launch."

GRAVITATIONAL

The violent merger of two black holes, as shown in this illustration, would give off gravitational waves, detected for the first time in 2016.

WAVE SPOTTING

HARNESSING A CHUNK OF SPACE BIGGER THAN OUR SOLAR SYSTEM HAS LET SCIENTISTS DETECT THESE ULTRALOW-FREQUENCY WAVES.

WORDS BY
PATRICIA S. DANIELS

ALL WORLDS, including our own planet, are bathed in light. Stars in their multitudes give off electromagnetic radiation, of which the visible spectrum is just a small part, and those waves travel at the speed of light throughout space. Over the past century scientists have realized that the universe is also awash in a different kind of radiation: gravitational waves, predicted by Albert Einstein in a 1915 paper. In 2015, detectors on Earth first registered these wobbles from a source 1.3 billion light-years away (one light-year equals 5.8 trillion miles). In 2023, using a "lens" as large as a galaxy, astronomers detected far fainter oscillations that may inform us about the birth of the cosmos itself.

Every moving object gives off gravitational waves; a bird flying through the air sends out its own tiny shivers in space. These waves are far too weak for us to detect. The ones we've spotted are produced when enormous masses collide violently, starting waves that spread outward from the event like ripples in a pond. Typically, this happens when two orbiting black holes or neutron stars (the extremely dense, compact remains of supernovae) spiral into each other. The waves could start from as long ago as the superviolent events of the big bang.

But gravitational waves are by their nature faint, and like other waves they lose energy as they travel. They are also enormous, as measured from the peak of one wave to the peak of the next (their frequency). Light waves are tiny, measured in nanometers, about the size of a small molecule. The shortest gravitational waves span hundreds of miles from peak to peak. The longest, which spread out from massive collisions in the early years of the universe, can be trillions of miles long.

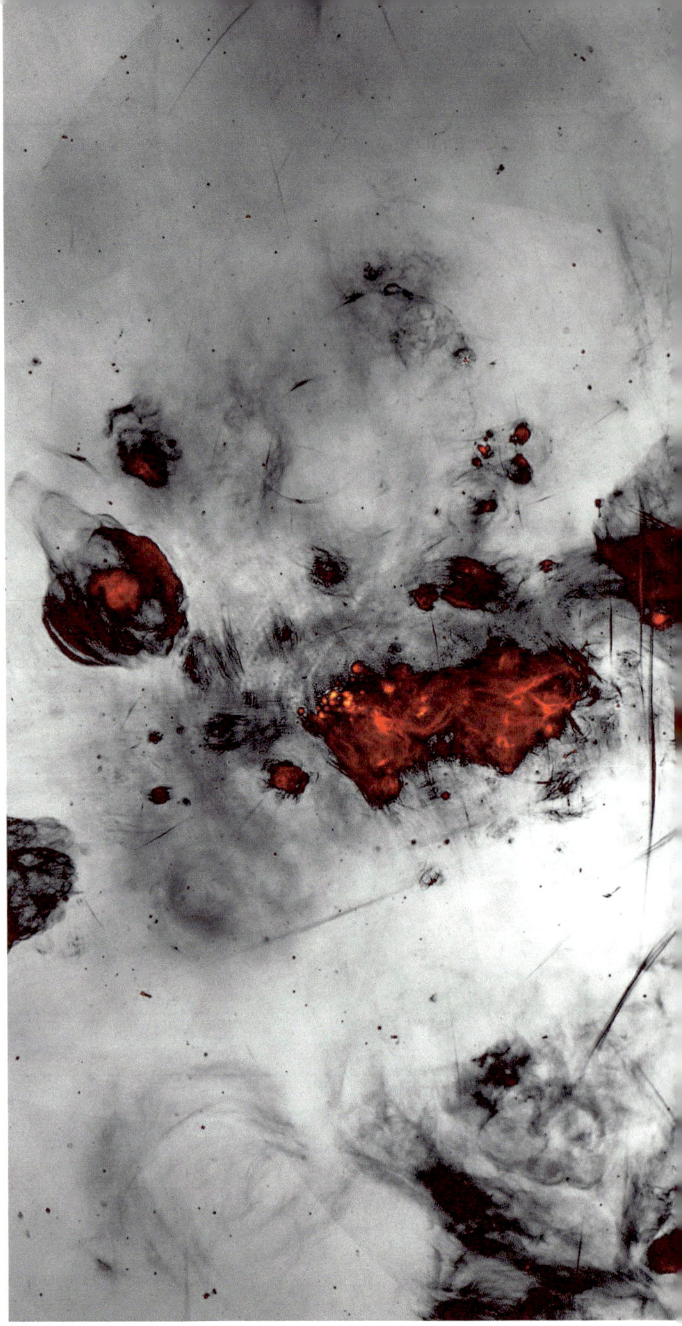

LOOKING AT WAVES WITH THE STARS

THE EARTHBOUND DETECTORS that spotted the 2015 waves have arms more than two miles long, but the gravitational ripples they were detecting shook them by less than the span of an atomic nucleus. To find the longest waves, then, astronomers knew they needed an instrument with a lens bigger than our solar system. Okay, they said, and built one from the stars.

Founded in 2007, the North American Nanohertz Observatory for Gravitational Waves (NANOGrav) project and its international partners collected pulsar observations from around our own Milky Way galaxy. Pulsars are spinning neutron stars: When a massive star explodes, it can collapse into a dense mass about 12.5 miles across, no larger than a typical city. One spoonful of its intensely packed material would weigh about one billion tons. As they spin, pulsars' intense magnetic fields channel radio waves outward like beams from a lighthouse. Some of those beams sweep across the Earth, *tick, tick, ticking* in our radio telescopes like distant, hyperaccurate clocks.

It is the very precision of pulsars that makes them suitable for registering gravitational waves. Scientists knew that when such a wave swept past a pulsar, it would shake the star just a little and disrupt the timing of its regular signal. Over 15 years, three radio telescopes on Earth tracked the signals from 68 pulsars, noting any tiny deviations in their regular flashes. The patient process paid off in 2023, when astronomers announced that

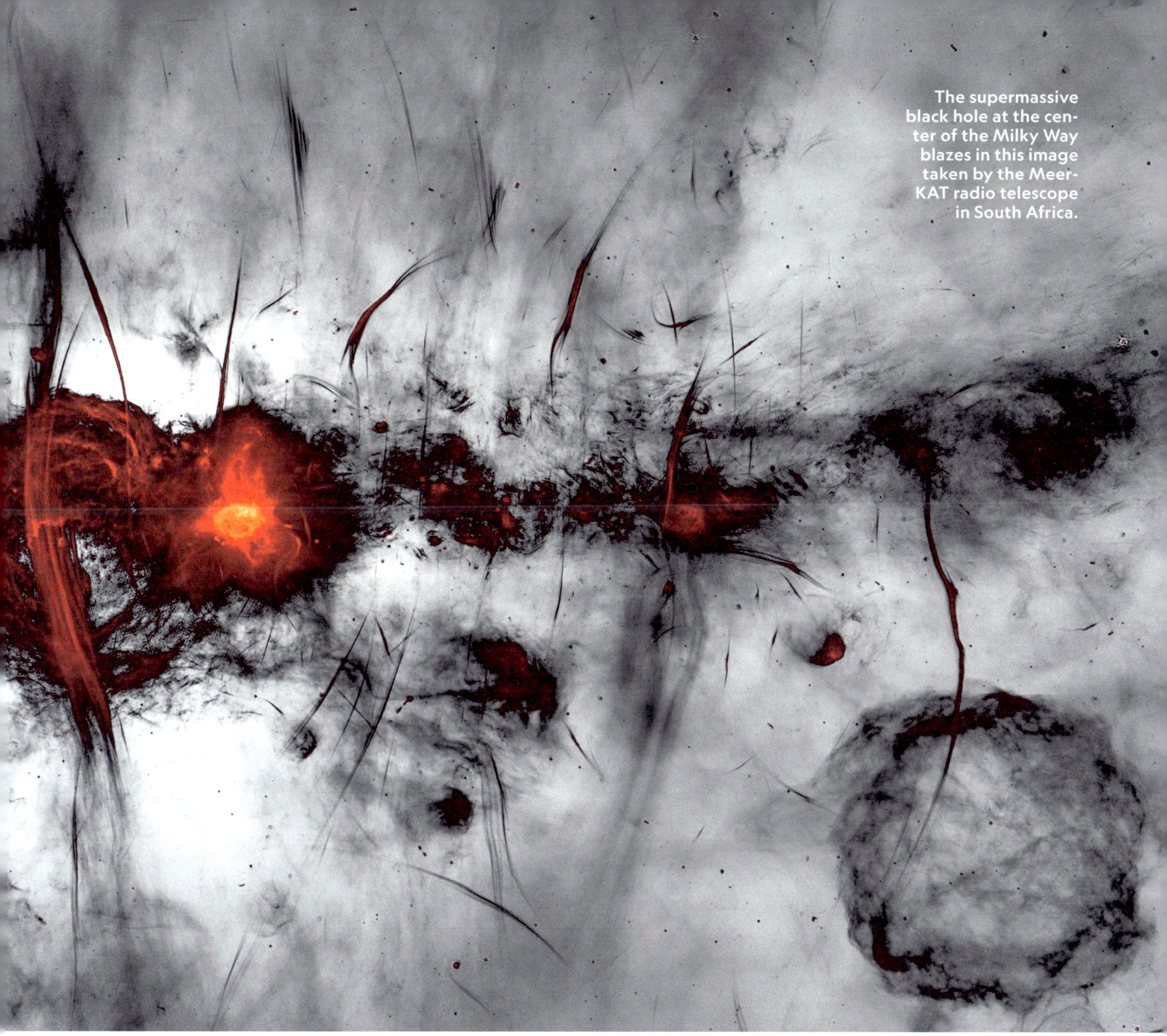

The supermassive black hole at the center of the Milky Way blazes in this image taken by the MeerKAT radio telescope in South Africa.

their galaxy-wide detector was in fact being shaken by low-frequency waves, a whole chorus of them in fact, humming a quiet song.

UNUSUAL SUSPECTS

BECAUSE IT TAKES A MASSIVE event to start such a wave rolling outward, the celestial choir must reflect titanic collisions or explosions. Astronomers can't say for certain yet what those events were, but they have several suspects. One possibility is the merger of supermassive black holes at the hearts of two galaxies. When two galaxies come close enough to be attracted to one another, they can begin a long, spiraling dance that flings out gravitational waves, ending as their centers merge into an even more massive black hole. Studying the hum from collisions near and far could help teach astronomers how galaxies have formed and changed over time.

The NANOGrav finding is particularly exciting for those who study the early cosmos. Some of the signals may come from cosmic collisions very early in galactic history. There's even a chance that waves passing through our detectors now originated with the big bang.

The early history of our cosmos is still filled with mysteries. According to radio astronomer Scott Ransom, "Combining these gravitational-wave results with studies of galaxy structure and evolution . . . will revolutionize our understanding of the history of our universe."

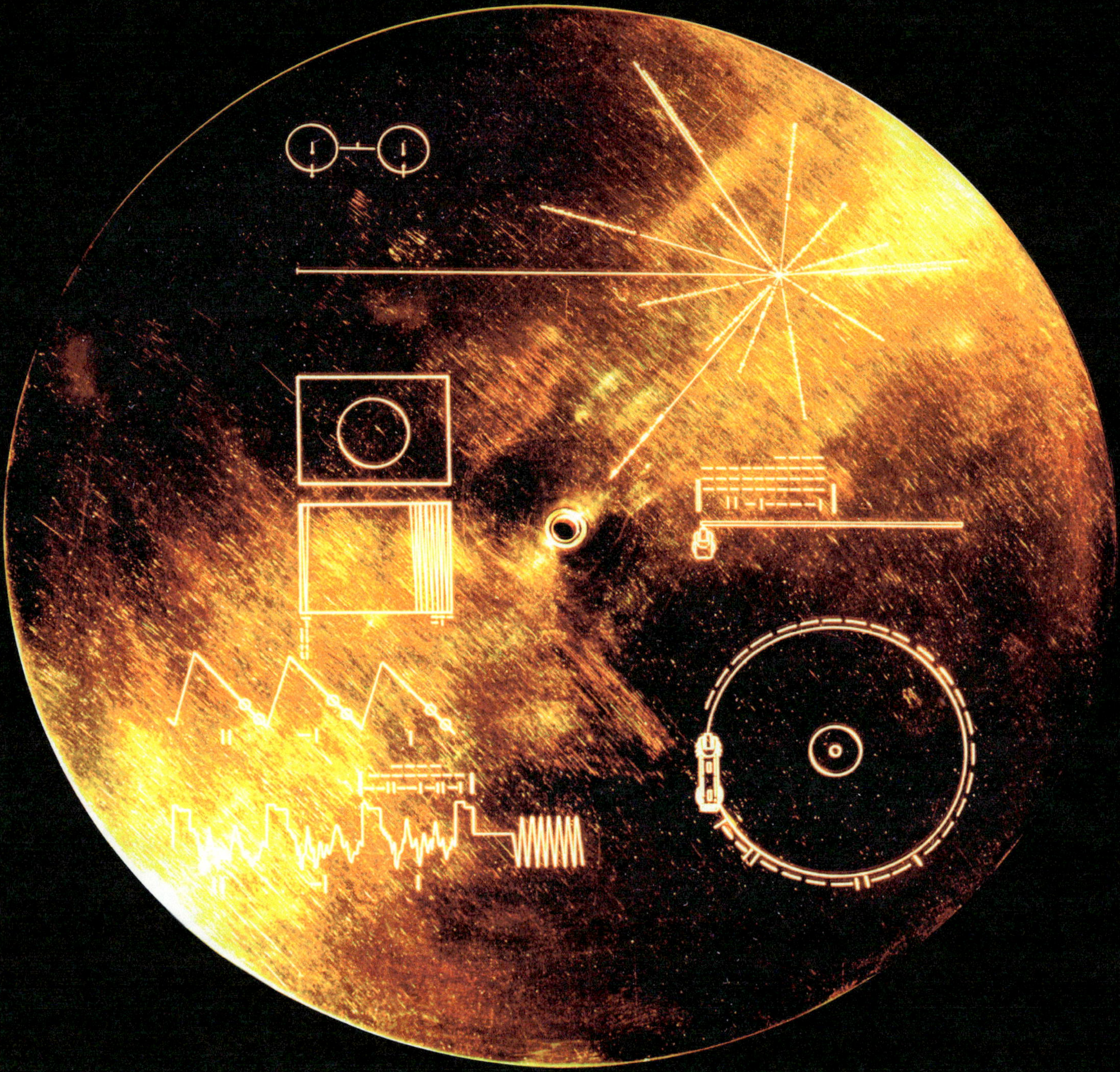

DEAR VOYAGERS,

YOUR BILLION-YEAR JOURNEY CARRIES TRUE LOVE.

WORDS BY **ANN DRUYAN**

Ann Druyan was creative director for *Sounds of Earth*, carried on the Voyager 1 and 2 spacecraft. The golden phonograph records (above) hold sounds and images of our planet.

YOU, THE FARTHEST objects we have ever touched, now venture beyond that place where the sun's wind gives way to roaring interstellar gales; far, far away—and yet I feel close to you. My own life attained lift-off when you did. I was 27 years old when you were in the last stages of your assembly. Carl Sagan and I had known each other as friends and colleagues for a couple of years. We fell in love in 1977 while collaborating on the message that you carry. At the same moment that you left Earth to discover and explore other worlds and to blaze our trail to the stars, we were launched on our own life trajectory. You have been on my mind ever since.

The golden discs affixed to both of you are rich in information, including our return address. The scientific hieroglyphic on them that resembles a burst of fireworks is actually a map of the frequencies of the thirteen nearest pulsars. We think that these rapidly rotating neutron stars—virtually inexhaustible natural beacons, each identifiable by their unique rate of spin—will point the way to our sun and its system of worlds.

Countless times I have tried to imagine that I am flying along with you at 38,000 miles an hour past gas giants and ice worlds, as you leave the shallows of the cosmic ocean, bullet-like, even under the punishing assault of cosmic rays.

I have fantasized your discovery by an extraterrestrial civilization. They reel in one of you derelict spacecraft, assessing your Model T technology, and pore over the symbols on your golden disc, as Champollion and Young once did in their efforts to decipher the texts of ancient Egypt. But your deeper message is hidden beneath. Someone figures this out and pries off the cover to find 27 pieces of music from the world's cultures, 118 images of life here, greetings in the languages of humans and animals, and an audio essay about life on Earth.

The crackle of thunder awaits them. A mother's first words to her newborn. A cricket's song. An hour of meditation by a young woman newly fallen in true love.

I was that woman.

Just days after Carl and I realized we would spend the rest of our lives together, I meditated for an hour while blindfolded, hooked up to every human monitoring device then known. The signals from my mind and heart—the most intimate message on the record—were translated into data and paired on the golden records with a strangely similar rasping, that of the most distant sound ever recorded back then: a pulsar.

Now, I am 70. Yet the feelings of that long-ago spring afternoon remain fresh and astounding. It was the first day of June, and I was searching for the piece of music that would honor China's 2,500-year-old continuous tradition. I did not know a single thing about Chinese music, so the search for its exemplar had been especially challenging. Once I felt sure I had found it, I left a message for Carl at his hotel in Tucson. We had never kissed or even joked about our feelings for each other, but when he returned my call, in the course of an exchange of about a dozen words, we had decided to marry.

> THE GOLDEN DISCS AFFIXED TO BOTH OF YOU ARE RICH IN INFORMATION, INCLUDING OUR RETURN ADDRESS.

And marry we did—in every way possible. Our families, our work, our hearts and minds and days and nights were blissful in their oneness, for the next two decades, until his death.

I and everyone I know will be dust for more than fifty thousand years before you near another star (unless some spacefarer should flag you down). But even if you are never claimed, you two have already taught us many things. Your dispatches from the outer solar system revealed new moons, geysers, volcanoes, subsurface oceans, hurricanes of a ferocity that would fracture our extended scale of categories, and even the very shape of our solar system as it moves through the Milky Way galaxy.

Under Carl's leadership, you gave us another gift: that lesson in humility known as the pale blue dot, a portrait of our one-pixel world taken from out by Neptune. It is a way to grasp our true circumstances that can pierce even the fiercest form of denial. To see it is to know that we all live on a tiny dot. I wonder how long the petty leaders and polluters can hold onto their delusions now that we've seen the picture you took and sent home to us, Voyager 1? From where you are now, Earth would be invisible.

For me you are more than machine. You are also an apt metaphor for Carl. In him, as in you, our science and humanity were united without conflict. Wonder and skepticism, imagination and rigor, ambition and inclusion, passion and reason, audacity and humility, precision and tenderness—none ever at the cost of the other—all combined to make him what he was, and you what you are.

From the 1970s through the 1980s, the two Voyager spacecraft beamed back to Earth images of (clockwise from top left) Jupiter, Uranus, Saturn's rings, and Neptune.

Our collective human memory can barely reach back ten thousand years. Generations of scientists have been reconstructing the past of our species, our planet, and our universe. Thanks to them, we have an idea of what happened nearly 14 billion years ago. Your projected shelf life measures on that same enormous timescale. You are expected to complete four to twenty trips around the galaxy with our music, our images, our messages—intact. Since you have exceeded your mission specifications in every other way, I favor the optimistic view of a five-billion-year-long shelf life for your message. That's likely longer than the future history of life on Earth. Your voices may still speak for us when our sun has become a red giant and all terrestrial remnants of our existence have been reduced to ash.

Even then, because of you, we can imagine that five thousand million years from now, our blues and our ragas, and a heart at its greatest fullness, will sing on.

Ad astra, my darlings.
My love and I are with you always,
Ann, July 10, 2019

MYSTERIES OF SPACE

PRODUCED BY
NATIONAL GEOGRAPHIC PARTNERS, LLC
1145 17th Street NW
Washington, DC 20036-4688 USA

Material from this bookazine was adapted from the following source: "NASA is preparing to explore alien worlds—by investigating Earth's dark corners," © 2023 Nadia Drake, *National Geographic,* October 2023.

ISSN 2160-7141

Published by Meredith Operations Corporation
225 Liberty Street • New York, NY 10281

Printed in the USA

Special thanks to McKenzie Baker, Patricia S. Daniels, Emily Shenk Flory, and Sharon Moore.

SUSTAINABLE FORESTRY INITIATIVE
Certified Sourcing
www.forests.org
SFI-00993

Printed on SFI-certified paper.

ILLUSTRATIONS CREDITS
Cover, X-ray: NASA/CXC/SAO; Optical: NASA/ESA/STScI; IR: NASA/ESA/CSA/STScI/Milisavljevic et al., NASA/JPL/CalTech; Image Processing: NASA/CXC/SAO/J. Schmidt and K. Arcand; 0–1, NASA, ESA, CSA, STScI, Webb ERO Production Team; 2–3, NASA, ESA, CSA, STScI; 4, Otis Imboden/National Geographic Image Collection; 5, NASA; 6, NASA/Bill Ingalls; 8, Dan Winters/National Geographic Image Collection; 9, Dan Winters/National Geographic Image Collection; 11, Dan Winters/National Geographic Image Collection; 12–13, Dan Winters/National Geographic Image Collection; 14–15, Dan Winters/National Geographic Image Collection; 16–17, Rebecca Hale/National Geographic Image Collection; 18–19, Dan Winters/National Geographic Image Collection; 20, Dan Winters/National Geographic Image Collection; 21, Dan Winters/National Geographic Image Collection; 22 (LO RT), Mattel, Inc.; 23 (LO LE), Mattel, Inc.; 23 (LO RT), Mattel, Inc.; 23 (LO CTR), Mattel, Inc.; 24 (UP LE), Dan Winters/National Geographic Image Collection; 24 (UP RT), Dan Winters/National Geographic Image Collection; 24 (LO LE), Dan Winters/National Geographic Image Collection; 25 (UP LE), Dan Winters/National Geographic Image Collection; 25 (UP RT), Caltech/Caltech Space Solar Power Project/Simulation: Cosmic Renaissance, John Wise, Tom Abel, Brian O'Shea, Michael Norman/Visualization: Ralf Kaehler and Tom Abel, Stanford University/KIPAC/SLAC; 26–27, Dan Winters/National Geographic Image Collection; 32–33, kittisun kittayacharoenpong/Moment/Getty Images; 35, Dan Winters/National Geographic Image Collection; 36, NASA/JPL/Space Science Institute; 38–39, Carsten Peter/National Geographic Image Collection; 40–41, Carsten Peter/National Geographic Image Collection; 44–45, Carsten Peter/National Geographic Image Collection; 46–47, Carsten Peter/National Geographic Image Collection; 48, Harris Corporation/TJT Photography; 50–51, REV Ocean/HACON;52, NASA/JPL-Caltech/Space Science Institute; 53, Chris Gunn/National Geographic Image Collection; 60–61, NASA/JPL-Caltech/MSSS; 63, Babak Tafreshi/National Geographic Image Collection; 65, Joseph DePasquale (STScI), Alyssa Pagan (STScI), Anton M. Koekemoer (STScI)/NASA, ESA, CSA, STScI; 66–67, NASA ESA, CSA, STSCI; 68, Chris Gunn/National Geographic Image Collection; 70, ESA/WEBB, NASA, CSA, M. ZAMANI (ESA/WEBB), PDRS4ALL ERS TEAM; 71, NASA, ESA, CSA, STSCI; 72–73, Stanford University/KIPAC/SLAC; 72 (UP LE), NASA, CSA, M. ZAMANI (ESA/WEBB), LEAH HUSTAK (STSCI); 74–75, NASA, ESA, KRUK, AND OTHERS; 75, NASA, ESA, KRUK, AND OTHERS; 76, Bettman/Contributor/Getty Images; 77, NASA, ESA, CSA, STScI, Janice Lee (STScI), Thomas Williams (Oxford), PHANGS Team; 78–79, NASA, ESA, CSA, STSCI; 80–81, NASA, ESA, AND THE HUBBLE HERITAGE TEAM (STSCI/AURA); 81 (UP LE), NASA/Yuchen Guo, Shardha Jogee, Steven F. Finkelstein, and the CEERS Collaboration; 81 (UP CTR), NASA/Yuchen Guo, Shardha Jogee, Steven F. Finkelstein, and the CEERS Collaboration; 81 (UP RT), NASA/Yuchen Guo, Shardha Jogee, Steven F. Finkelstein, and the CEERS Collaboration; 81 (LO LE), NASA/Yuchen Guo, Shardha Jogee, Steven F. Finkelstein, and the CEERS Collaboration; 81 (LO CTR), NASA/Yuchen Guo, Shardha Jogee, Steven F. Finkelstein, and the CEERS Collaboration; 81 (LO RT), NASA/Yuchen Guo, Shardha Jogee, Steven F. Finkelstein, and the CEERS Collaboration; 82–83, Chris Gunn/National Geographic Image Collection; 84–85, Chris Gunn/National Geographic Image Collection; 86–87, Chris Gunn/National Geographic Image Collection; 88–89, Nicolle R. Fuller/Science Source; 90–91, South African Radio Astronomy Observatory/National Geographic Image Collection; 92–93, NASA/digitaleye/J Marshall - Tribaleye Images/Alamy Stock Photo; 95 (UP RT), NASA/JPL; 95 (UP LE), NASA; 95 (LO LE), NASA/JPL; 95 (LO RT), NASA/JPL; 97, NASA, ESA, CSA, STScI, Danny Milisavljevic (Purdue University), Ilse De Looze (UGent), Tea Temim (Princeton University).

Made in the USA
Monee, IL
05 November 2026

67387481R00059

The expanding supernova remnant Cassiopeia A appears in detail in an image from Webb.